Tales of the
Night Sky

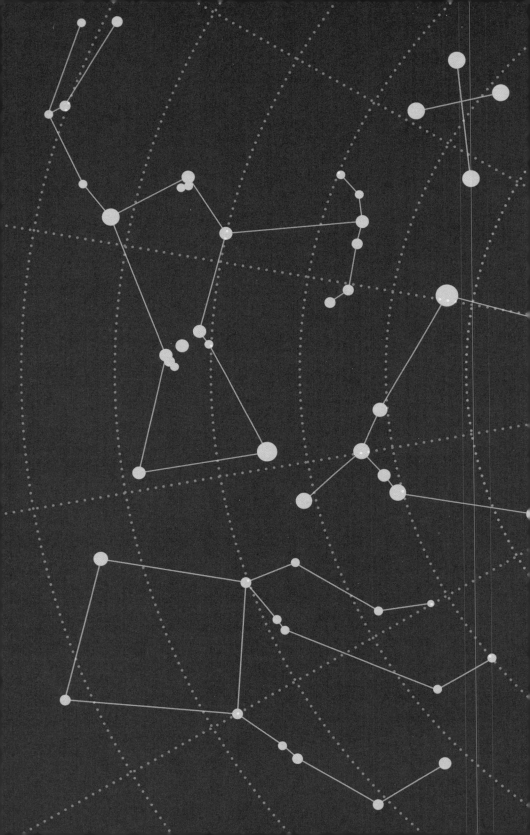

Tales of the Night Sky

Revealing the MYTHOLOGIES and FOLKLORE behind the CONSTELLATIONS

Robin Kerrod

WELLFLEET
PRESS

Inspiring | Educating | Creating | Entertaining

Brimming with creative inspiration, how-to projects, and useful
information to enrich your everyday life, Quarto Knows is a favourite
destination for those pursuing their interests and passions. Visit our
site and dig deeper with our books into your area of interest:
Quarto Creates, Quarto Cooks, Quarto Homes, Quarto Lives,
Quarto Drives, Quarto Explores, Quarto Gifts, or Quarto Kids.

Copyright © 2020 Quarto Publishing plc,
an imprint of The Quarto Group
The Old Brewery, 6 Blundell Street
London N7 9BH

This edition published in 2020 by Wellfleet Press,
an imprint of The Quarto Group,
142 West 36th Street, 4th Floor,
New York, NY 10018, USA
T (212) 779-4972 F (212) 779-6058
www.QuartoKnows.com

Wellfleet titles are also available at discount for retail, wholesale,
promotional, and bulk purchase. For details, contact the Special
Sales Manager by email at specialsales@quarto.com or by mail at
The Quarto Group, Attn: Special Sales Manager, 100 Cummings
Center Suite 265D, Beverly, MA 01915 USA.

Library of Congress Control Number: 2020936901

10 9 8 7 6 5 4 3 2 1

ISBN: 978-1-57715-228-6

Printed in Singapore

For entertainment purposes only. Do not attempt any spell,
recipe, procedure, or prescription in this book. The author,
publisher, packager, manufacturer, distributor, and their
collective agents waive all liability for the reader's use or
application of any of the text herein.

Contents

Introduction

Dome of the Heavens

When you go stargazing on a clear, dark night, the star-studded heavens form a vast, dark dome over your head. It is the same everywhere on Earth. Our planet is surrounded by a limitless celestial sphere, and the stars appear to be fixed to the inside of this sphere. As time goes by, the stars wheel over your head and the sphere then seems to be spinning around the Earth.

At first glance, there is little to distinguish individual stars, and they seem to be scattered haphazardly around the dark dome of the heavens. But it soon becomes evident that the stars are not all alike. Some are so bright they stand out like beacons; others are so dim you can scarcely make them out. In your mind's eye, you can group together some of the bright stars to make patterns.

Night after night, you will be able to find these same patterns in the sky. Even though the stars appear to wheel overhead every night, they move together bodily—they do not change their relative positions in their patterns, which we call the constellations.

The stars seem to be fixed in position inside the celestial sphere. This is why they are often called the fixed stars. But there seem to be a few notable exceptions to this general observation: occasionally, five bright objects can be found wandering around the celestial sphere among the fixed stars in the constellations.

But appearances can be deceiving. Really, we know that there is no great enveloping dark celestial sphere surrounding the Earth. The darkness of the night sky is the profound blackness of empty space extending for distances so vast as to be beyond our human comprehension. The tiny pinpricks of light visible as stars are, in reality, huge globes of incandescent gas that pour forth enormous energy into space as light, heat, and other forms of radiation. They are distant Suns.

As for the wandering objects, they are not stars at all, but much smaller, closer bodies that we call the planets.

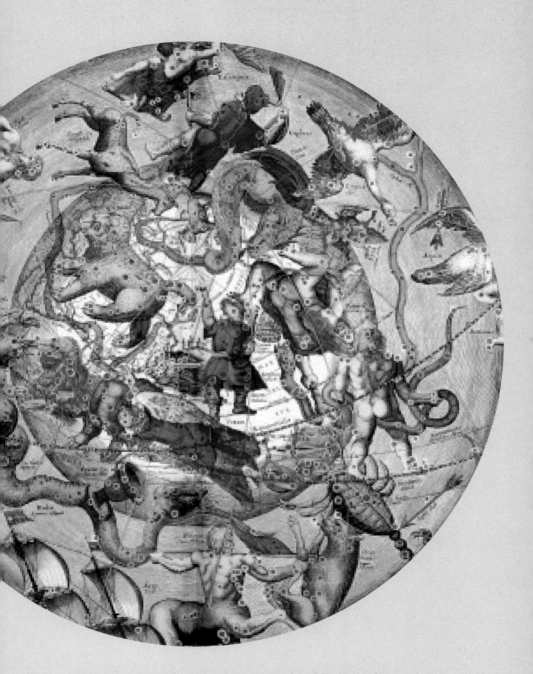

The constellations of the northern hemisphere, as pictured by Cellarius (1708).

The Universe

Two thousand years ago, astronomers believed that there was a celestial sphere surrounding Earth, and that Earth was the center of the universe. They also thought that the Sun, Moon, and planets all circled around Earth, as did the stars fixed inside the celestial sphere. This was the classic Greek concept of the universe, elaborated by the last great Greek astronomer Ptolemy of Alexandria in around 150 C.E. This era was marked by the fabulous, fantastic tales of Greek mythology. These stories told of the epic adventures and wondrous deeds of gods and goddesses, heroes and heroines, fantastic creatures like centaurs, and monsters like Gorgons.

The Universe
Ancient stories

In ancient Greece, astronomy and mythology intermingled, and the heavens became the immortal resting place for a host of mythological characters that were embodied in the constellations. Although we have inherited most of the Greek constellations, we know them today not by their original Greek names, but by their Latin ones.

Between two rivers

The Greeks inherited many of the constellations from earlier times—predominantly from Mesopotamia. The Greeks gave the name Mesopotamia, meaning "between two rivers," to the region between the rivers Tigris and Euphrates in what is modern Syria and Iraq. This was where the Sumerians and Babylonians built up the first great civilizations in the Middle East, and where writing and the wheel were invented around 3000–3500 B.C.E. The first writing took the form of picture symbols, or pictographs. In pictographs on decorated pottery, in carvings and on seals, three figures were common—the bull, the lion, and the scorpion. These three figures were pictured in the sky in the earliest zodiacal constellations—the constellations the Sun passes through each year. They were the forerunners of Taurus, Leo, and Scorpius.

The heavenly triad

Later artworks showed other animals and gods, some clearly identified with heavenly bodies. Symbols of the Sun, Moon, and Venus—the three brightest heavenly bodies—became a recurrent theme. This triad was handed down to the Babylonians, where Venus was Ishtar, the "queen of heaven and whore of Babylon." Fine representations of these and the zodiacal constellations have survived on boundary stones of Babylonian times, dating from about 1300 B.C.E.

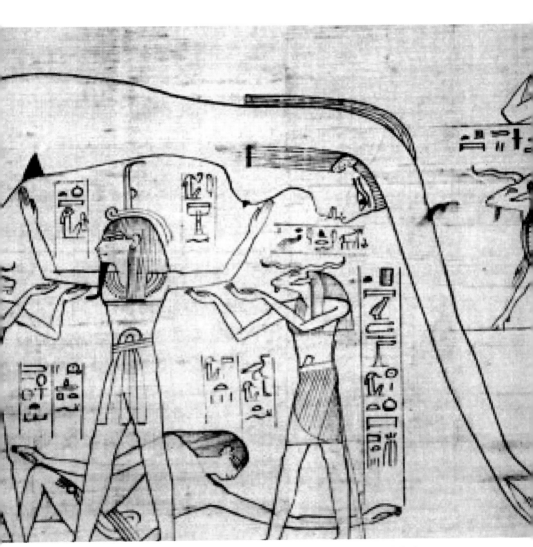

In ancient Egyptian mythology, the starry body of the sky goddess, Nut formed the heavens.

Ptolemy's Universe

An examination of planetary motion

By Ptolemy's time, astronomers were recording the positions of the wandering stars—the planets—in the sky with some accuracy. And they realized that there was something odd about the motions of the planets. If they indeed circled around Earth as was believed, they should always travel in the same direction—toward the west, like the Sun. But often a planet could be seen to backtrack in the sky, moving eastward for some time before resuming its usual westward course.

Ptolemy developed a system to try to account for this occasional backward motion. He said that a planet moved in a little circle (epicycle) around a point (deferent) that moved in a great circle around the Sun. This did not work very well either, and so over the years, further epicycles were introduced until the system became impossibly complex.

Into and out of the Dark Ages

The Ptolemaic, Earth-centered view was accepted for nearly 1,400 years. In Europe, with the decline of Greco-Roman civilization by about the fifth century, Europe slipped into a period of general cultural stagnation when much of the knowledge of the ancient world was either lost or forgotten.

Fortunately, astronomy continued to thrive elsewhere, particularly in Arabia. One of the triggers was the translation into Arabic of Ptolemy's seminal work the *Almagest*, ca. 820 C.E. This inspired generations of Arab astronomers until 1428, when Ulugh Beigh established at Samarkand the finest observatory the world had ever seen.

It was also in the 1400s that the great rebirth of learning we call the Renaissance was getting underway in Europe. Philosophers and scholars began questioning and investigating age-old beliefs. In astronomy, a startling breakthrough came from an unlikely source—a cleric and physician named Nicolaus Copernicus.

Copernicus had a passion for astronomy. He came to realize that Ptolemy's concept of the universe was wrong. The odd movements of the planets could be explained simply if the Sun, and not Earth, was the center of the universe. The planets circled around the Sun, and so did Earth. Earth was merely another planet.

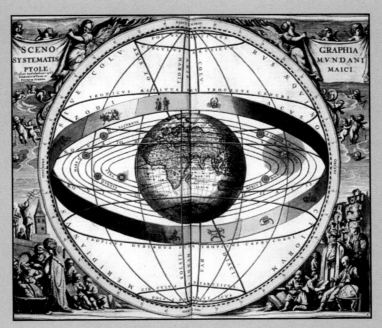

Ptolemy's idea of an Earth-centered universe, set out in an encyclopedic work known as Almagest *(The Greatest).*

The Sun was at the center of Copernicus' universe.

Galileo explains the sights seen through his telescope.

The Universe

De Revolutionibus

A solar system is proposed

Copernicus was not the first person to put forward the idea of a universe centered on the Sun. The Greek philosopher Aristarchus had done so around 200 B.C.E. But no one listened because his idea placed Earth in an inferior position in the universe.

Copernicus knew that his concept of a solar system would upset the Church, which at that time virtually dictated all opinions within society. Anything that went against Church beliefs was tantamount to heresy, and punishable by excommunication, torture, and even death. So it was not until Copernicus was on his deathbed in 1543 that he published his ideas in *De Revolutionibus Orbium Coelestium.*

The evidence mounts

The idea of a solar system took hold only slowly but, in 1609, two events consigned Ptolemy's Earth-centered view to history. Johannes Kepler in Germany calculated that the planets travel around the Sun, not in circles but in elliptical (oval) orbits, which matched planetary observations exactly.

Also in 1609, Galileo Galilei in Italy turned a telescope on the heavens and saw sights no one had ever seen before. He saw mountains on the Moon and moons circling Jupiter. He also noticed that Venus showed phases, which could only happen if the planet circled around the Sun. This provided convincing evidence of a solar system.

The expanding universe

As telescopes became more powerful, astronomers began to probe deeper into space. In 1781, William Herschel spotted a new planet, Uranus. It proved to be twice as far from Earth as Saturn, the most distant planet known to the ancients. At a stroke, the size of the solar system had doubled. It expanded farther when Neptune was discovered in 1846, and Pluto as recently as 1930.

By the 1930s, the solar system was recognized as a family of bodies centered on the Sun. The Sun is one of billions of stars that belong to a great galaxy and many such galaxies form clusters to make the universe. The universe is not a static arrangement of galaxies scattered through space— the galaxies are rushing away from one another. Astronomers believe that an explosion took place about 15 billion years ago. This Big Bang created the universe and set it expanding. Current evidence suggests that it is going to expand forever.

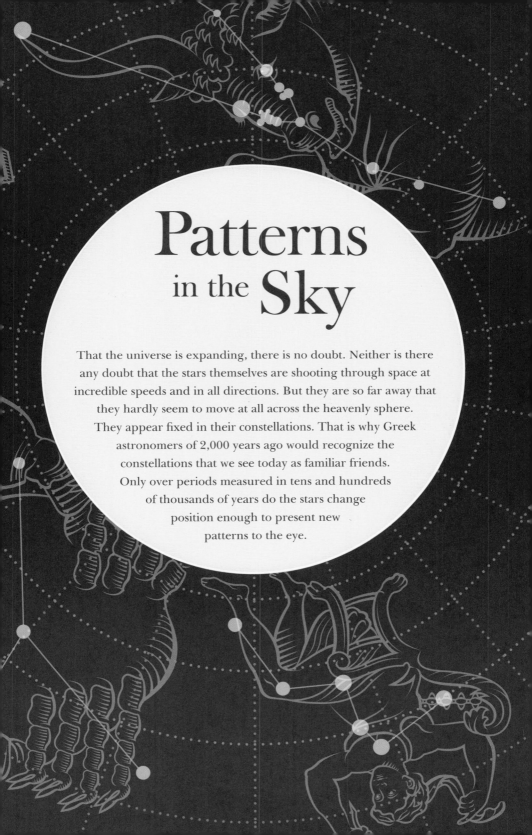

Patterns
in the Sky

That the universe is expanding, there is no doubt. Neither is there
any doubt that the stars themselves are shooting through space at
incredible speeds and in all directions. But they are so far away that
they hardly seem to move at all across the heavenly sphere.
They appear fixed in their constellations. That is why Greek
astronomers of 2,000 years ago would recognize the
constellations that we see today as familiar friends.
Only over periods measured in tens and hundreds
of thousands of years do the stars change
position enough to present new
patterns to the eye.

An ancient print from 1515 showing northern constellations.

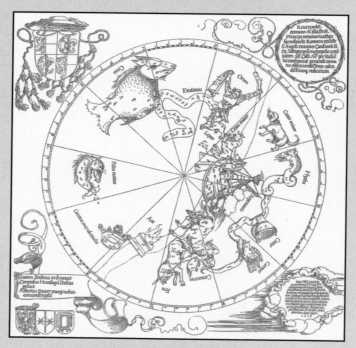

An ancient print from 1515 showing southern constellations.

Patterns in the Sky

Patterns in the Sky
Identifying the constellations

According to Ptolemy, the Greeks recognized forty-eight constellations. They named them after figures they thought the patterns of bright stars formed in the sky. In a few cases, we can see what they saw so long ago. One constellation does look rather like a flying swan, another like a lion, and another like a scorpion. We know these and the other constellations by the Latin versions of the names the Greeks gave them. So the swan is Cygnus, the lion is Leo, and the scorpion is Scorpius. In most instances, however, immense imagination is needed to associate a pattern of stars in the sky with a specific constellation name.

On with the new

Forty new constellations have been added since Ptolemy's time, making a total of eighty-eight in all. It might be expected that Arab astronomers would have introduced some since they, essentially, kept astronomy going through the Dark Ages. But they did not. Their main legacy has been in the names of many of the bright stars in the constellations, such as Betelgeuse in Orion, Algol in Perseus, Aldebaran in Taurus, and the beautiful Zubenelgenubi in Libra.

Three astronomers were mainly responsible for the additional constellations—in Germany, Johann Bayer in 1603 and Johann Hevelius in 1690; and in France, Nicolas Lacaille in 1752. It is to Bayer that we are also indebted for our system of identifying stars in a constellation by a letter of the Greek alphabet. They are graded in approximate order of brightness from alpha (brightest) onward.

The Celestial Sphere

Pinpointing the stars

As we have noted, it seems as if the stars—the constellations—are stuck on the inside of a great celestial sphere surrounding Earth. And this great sphere seems to spin around Earth, making the stars travel across the night sky from east to west.

But just as the celestial sphere is an illusion, so is its motion. The heavens are not spinning, it is Earth that is in motion. Earth spins around in space once a day and it is this that makes the Sun appear to arc across the sky by day, and the stars to wheel across the sky at night. And Earth spins around in the opposite direction from the Sun and the stars—from west to east.

The celestial sphere appears to spin around on an axis through the north and south celestial poles. These are points on the celestial sphere directly above Earth's North and South Poles. The celestial equator divides the sphere in two—into northern and southern celestial hemispheres. It is a projection on the celestial sphere of Earth's Equator, which divides Earth into the Northern and Southern Hemispheres. For convenience, we often divide the constellations in two according to the hemisphere in which they appear— northern or southern (see pages 22–23).

The concept of the celestial sphere may be ancient but it still has a role to play in modern astronomy. From an observer's point of view, it reflects exactly what we see in the heavens. In particular, it provides a simple means of pinpointing the stars in the sky. By using the geometry of a sphere, astronomers locate a star using a grid system analogous to the latitude and longitude system geographers use for locating a place on Earth. They pinpoint a star on the celestial sphere by its celestial latitude and longitude. The celestial latitude (called declination) is the angular distance of the star north or south of the celestial equator, just as latitude on Earth is the angular distance north or south of the Equator.

Celestial longitude is the distance around the celestial sphere from a fixed point (the First Point in Aries) just as longitude on Earth is the distance along the Equator from a fixed point (the Greenwich Meridian). Celestial longitude is known as right ascension (RA).

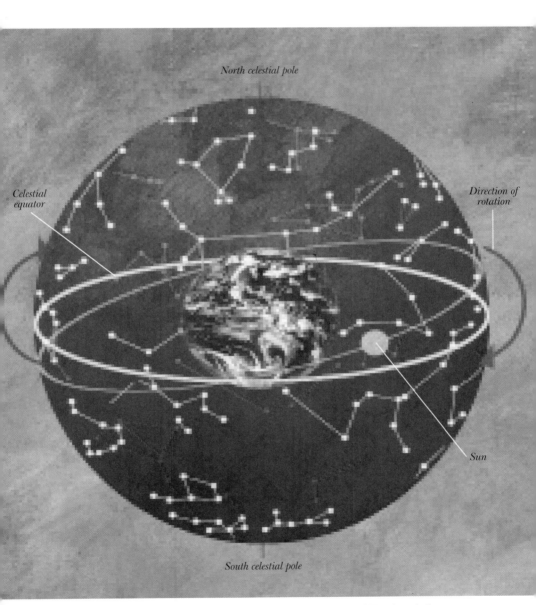

North celestial pole

Celestial
equator

Direction of
rotation

Sun

South celestial pole

The imaginary celestial sphere, which seemingly rotates around Earth each day.

Northern Constellations

The constellations of the northern celestial hemisphere

The darker area shown on the celestial sphere marks the Milky Way.

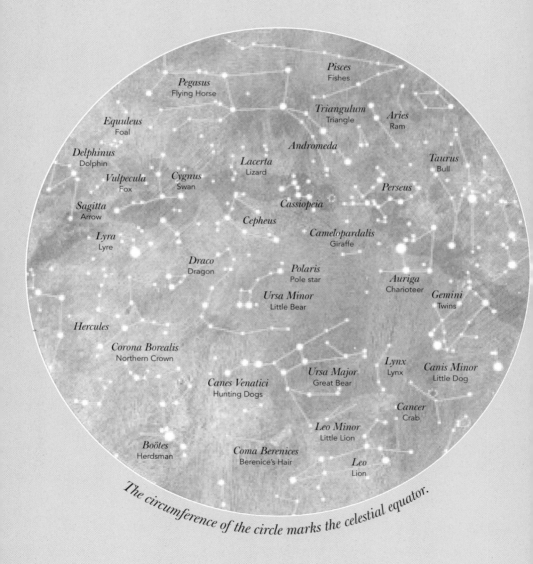

Pisces
Fishes

Pegasus
Flying Horse

Triangulum
Triangle

Aries
Ram

Equuleus
Foal

Andromeda

Taurus
Bull

Delphinus
Dolphin

Lacerta
Lizard

Vulpecula
Fox

Cygnus
Swan

Perseus

Sagitta
Arrow

Cassiopeia

Cepheus

Lyra
Lyre

Camelopardalis
Giraffe

Draco
Dragon

Polaris
Pole star

Auriga
Charioteer

Ursa Minor
Little Bear

Gemini
Twins

Hercules

Corona Borealis
Northern Crown

Lynx
Lynx

Canis Minor
Little Dog

Canes Venatici
Hunting Dogs

Ursa Major
Great Bear

Cancer
Crab

Leo Minor
Little Lion

Boötes
Herdsman

Coma Berenices
Berenice's Hair

Leo
Lion

The circumference of the circle marks the celestial equator.

Southern Constellations

The constellations of the southern celestial hemisphere

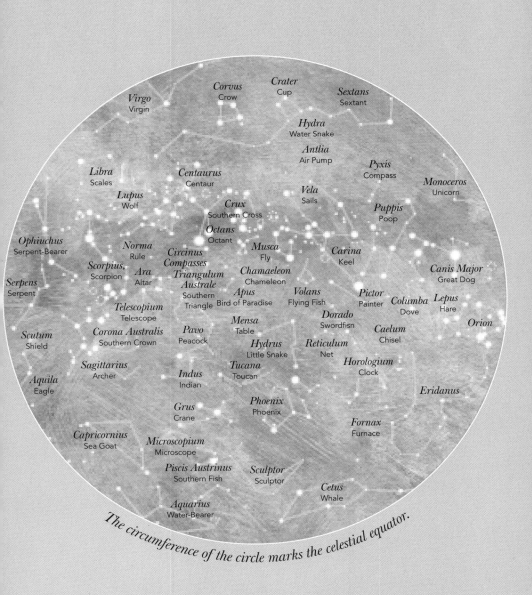

Virgo — Virgin
Corvus — Crow
Crater — Cup
Sextans — Sextant
Hydra — Water Snake
Antlia — Air Pump
Pyxis — Compass
Monoceros — Unicorn
Libra — Scales
Centaurus — Centaur
Vela — Sails
Lupus — Wolf
Crux — Southern Cross
Puppis — Poop
Octans — Octant
Ophiuchus — Serpent-Bearer
Norma — Rule
Circinus — Compasses
Musca — Fly
Carina — Keel
Scorpius, — Scorpion
Ara — Altar
Triangulum Australe — Southern Triangle
Chamaeleon — Chameleon
Canis Major — Great Dog
Serpens — Serpent
Apus — Bird of Paradise
Volans — Flying Fish
Pictor — Painter
Columba — Dove
Lepus — Hare
Telescopium — Telescope
Dorado — Swordfisn
Caelum — Chisel
Orion
Scutum — Shield
Corona Australis — Southern Crown
Pavo — Peacock
Mensa — Table
Reticulum — Net
Sagittarius — Archer
Hydrus — Little Snake
Horologium — Clock
Aquila — Eagle
Indus — Indian
Tucana — Toucan
Eridanus
Grus — Crane
Phoenix — Phoenix
Fornax — Furnace
Capricornius — Sea Goat
Microscopium — Microscope
Piscis Austrinus — Southern Fish
Sculptor — Sculptor
Cetus — Whale
Aquarius — Water-Bearer

The circumference of the circle marks the celestial equator.

What You Can See

Stargazing at night

Which constellations will you see when you go stargazing? That depends on many factors—where you are on Earth, the time of night, and also the time of year. Because Earth is spherical, people viewing from different latitudes will have different views of the celestial sphere—of the constellations.

If you stargaze from Toronto, Ontario, in Canada, high in the Northern Hemisphere, you will see different constellations than will fellow stargazers in Cape Town, South Africa, deep in the Southern Hemisphere. You will be able to spy the Big Dipper every night, but you will never get a glimpse of Crux, the Southern Cross (and vice versa, of course).

Telling time

Which constellations you see also depends on the time of night that you are viewing. Because the Earth is spinning, the stars move across the sky all the time. Some constellations rise in the east as others disappear beneath the western horizon.

Different constellations also appear and disappear as the months go by. This is a result of Earth traveling in its orbit around the Sun. Every day, Earth moves a little farther along its orbit, and every night when you stargaze you look out at a slightly different part of the celestial sphere. This difference soon becomes noticeable, and after three months—every season—the skies look quite different.

This leads us to look at the spring, summer, fall, and winter constellations. In the Northern Hemisphere, Orion is a spectacular winter constellation, while the Square of Pegasus warns that fall is on the way. In the Southern Hemisphere, the seasons are reversed. So Orion is a feature of summer skies, while Pegasus promises spring.

As a rule

To sum up, an observer in the Northern Hemisphere, if he or she remained stargazing all year, would see all the constellations of the northern celestial hemisphere and some of the southern celestial hemisphere. In a similar way, an observer in the Southern Hemisphere would be able to see all the southern constellations and some of the northern ones. In theory, the best place for stargazing is on the Equator, where all the constellations would be visible at some time of the year.

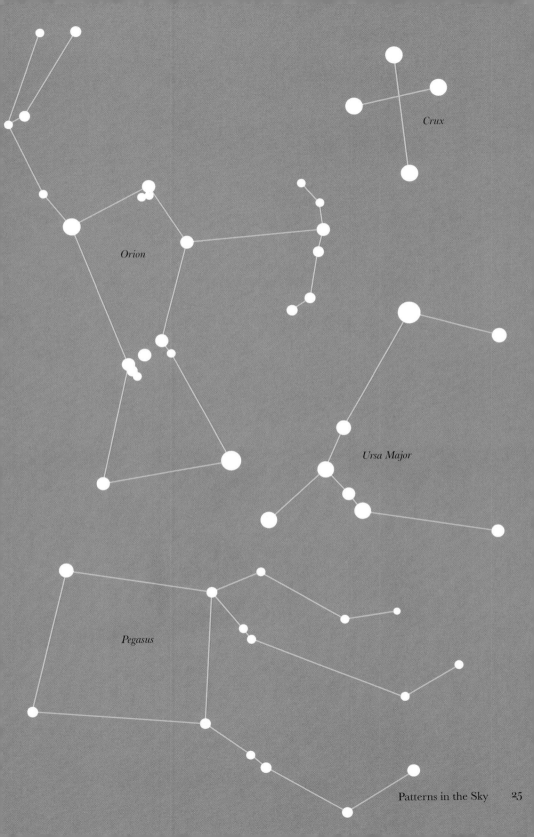

Crux

Orion

Ursa Major

Pegasus

Angels display the horoscope of Rudolf II, who became Holy Roman Emperor in 1576.

Constellations of the Zodiac

Where astronomy meets astrology

Earth orbits the Sun, making one complete journey every 365¼ days. But from our viewpoint on Earth, it appears that the Sun travels once around the heavens every year. We call the path of the Sun around the celestial sphere the ecliptic. The Sun follows the same path every year against the background of stars. The Moon and the planets are also always found in the sky quite close to the ecliptic, within an imaginary band in the heavens called the zodiac.

The constellations the zodiac passes through are known as the constellations of the zodiac. "Zodiac" roughly translated means "circle of animals," referring to the fact that most of the constellations are named after animals, such as Leo (lion) and Scorpius (scorpion).

The believers

Because the Sun and planets were always found among the zodiac constellations, ancient astronomers considered that the zodiac must have special significance. And a belief grew up that somehow the positions of the Sun and planets within the constellations affected people's characters and the lives they led. This belief became known as astrology. It thrived for more than 2,000 years, from Babylonian times to about the seventeenth century. Most rulers had a band of astrologers to advise them when to make important decisions. And the astrologers studied the heavens for propitious signs. It was the study of the stars and planets by astrologers that provided the springboard for astronomy, the true scientific study of the heavens. And even in our modern age, astrology has its adherents.

There is, however, no scientific basis for astrology. And there are powerful arguments against it. Astrologers base their belief on the position of the Sun among twelve constellations or "signs" of the zodiac—Aries, Taurus, Gemini, Cancer, Leo, Virgo, Libra, Scorpius, Sagittarius, Capricorn, Aquarius, and Pisces. But in fact there are thirteen constellations of the zodiac: the Sun passes through Ophiuchus on its way between Scorpius and Sagittarius.

Also, most modern day astrologers still assume that the Sun passes through each constellation at the same time as it did two millennia ago; but it doesn't. Because of precession—the slight gyration of the Earth's axis in space—the Sun passes through the constellations nearly a month earlier than it did in Roman times. This puts all the star signs out.

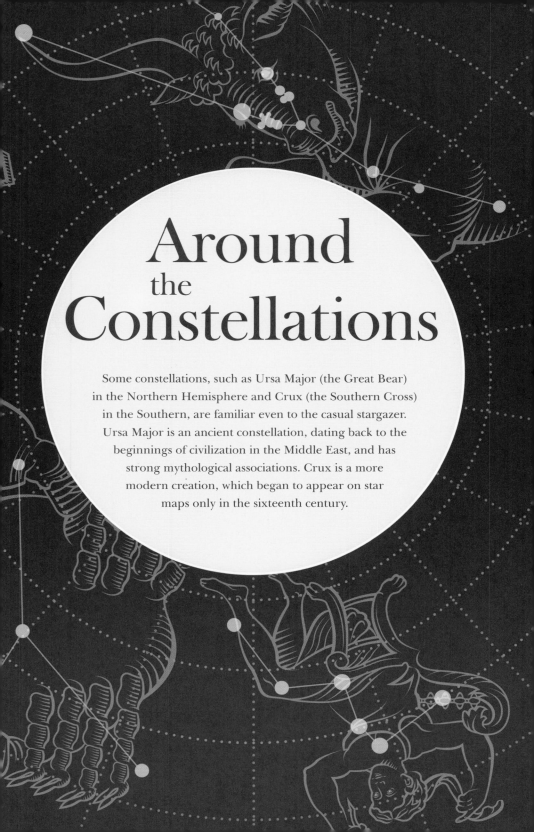

Around the Constellations

Some constellations, such as Ursa Major (the Great Bear) in the Northern Hemisphere and Crux (the Southern Cross) in the Southern, are familiar even to the casual stargazer. Ursa Major is an ancient constellation, dating back to the beginnings of civilization in the Middle East, and has strong mythological associations. Crux is a more modern creation, which began to appear on star maps only in the sixteenth century.

Around the Constellations
The star patterns that follow

Most of the thirty-three constellations included in this chapter date back to ancient times and are steeped in mythology. We look at some of the myths behind the figures traced by these star patterns and mention a few observational highlights. The four brightest stars of each constellation are indicated by the Greek letters Alpha (α), Beta (β), Gamma (γ), and Delta (δ). If other stars are mentioned in the text, they are also identified by their Greek letters. A few stars are named. Most of the names come from the Arabic, dating from the Middle Ages, when Arab astronomers led the world. Familiar examples are Betelgeuse and Rigel, the two brightest stars in Orion, and Algol, the famous variable star in Perseus. There are Greek names, such as Sirius, the brightest star in the sky, and Latin ones such as Bellatrix, another bright star in Orion.

Mention is often made of magnitude, which describes the brightness of a star. In naked-eye astronomy, astronomers grade the stars in six magnitudes of brightness as we see them in the sky. (We call this their apparent magnitude). The brightest ones are first magnitude, the faintest ones sixth magnitude; the others have magnitudes in between. This magnitude scale is extended beyond six to cover the brightness of much fainter stars, visible only in telescopes, and into negative values to express the brightness of exceptionally brilliant objects.

Twelve of the constellations featured are those of the zodiac, which provide the backcloth for the Sun's annual voyage around the celestial sphere. These are the constellations, or star signs that astrologers revere and that, they say, shape our character and dictate our destiny.

The stories behind the constellations are invariably fascinating. Here, we find a wronged maiden turned into a bear by a jealous wife—there, a hero slaying a snake-headed ogre. All fantastic life is here in these stirring, often racy tales.

Features of the constellation spreads that follow:

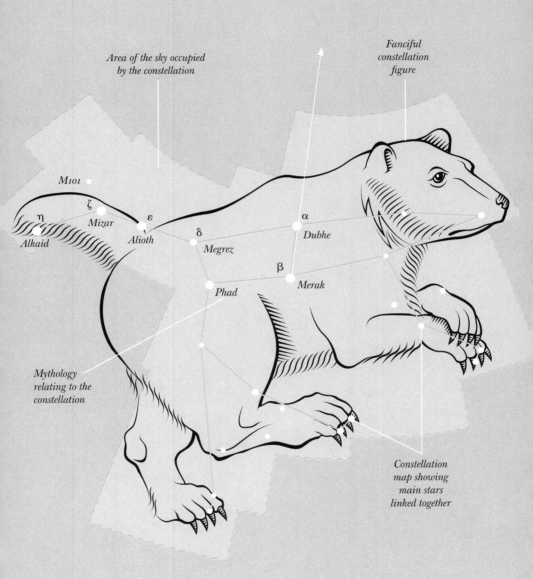

Area of the sky occupied
by the constellation

Fanciful
constellation
figure

M101

ζ

η

Mizar

ε

Alioth

Alkaid

δ

Megrez

α

Dubhe

β

Phad

Merak

Mythology
relating to the
constellation

Constellation
map showing
main stars
linked together

Andromeda
The Chained Lady

The constellation Andromeda is found in far northern skies, adjoining Perseus and Cassiopeia. It is often called the chained lady, for it was pictured on ancient star maps as a beautiful and tragic figure of a woman, who was bound by chains to rocks on the seashore. How did the fair Andromeda find herself in this predicament?

Andromeda was the daughter of King Cepheus and Queen Cassiopeia (see pages 50–51). The queen had offended some beautiful sea nymphs, called the Nereids, by claiming to be more beautiful than they. As retribution for this affront to his charges, the sea god, Poseidon sent a monster, Cetus (see pages 54–55), to wreak havoc along the shores of King Cepheus's kingdom.

In desperation, Cepheus consulted an oracle to find out if there was some way that he could put an end to the destruction the monster was causing. He was told that he had to give the monster a living sacrifice—his daughter—which is how Andromeda came to be chained to the rocks.

As the terrified Andromeda waited to be devoured, the hero Perseus happened along and asked who she was and why she was in this pitiful state. At first, improbably shy, Andromeda said nothing, but gradually she told Perseus everything, ending with an ear-splitting scream as she saw Cetus approaching. Perseus, fresh from killing the Gorgon Medusa, still had her head in his hand. He turned the Gorgon's face to Cetus, which promptly turned to stone and sank. As his reward, Perseus claimed Andromeda's hand in marriage and she later bore him six children.

The great spiral

Andromeda is easy to locate as it is joined to Pegasus, one of the most unmistakable of all constellations owing to its square of four stars. Visually, the individual stars in Andromeda are not particularly noteworthy. The most interesting object by far is a blurry patch, found roughly midway between the Square of Pegasus and the W-shape of Cassiopeia. This patch resembles a nebula or gas cloud, but in fact it is a distant star island, or galaxy. It is often called the Great Spiral because of its structure, which shows up when viewed through a powerful telescope.

The Andromeda Galaxy is the most distant object visible in the heavens with the naked eye. When you see it, you are looking across vast distances of space—its light has been traveling toward us for more than two million years.

Alpha marks the head of the helpless maiden, chained to the rocks to await her fate.

γ

υ

M31

β

π

δ

ε

α
Alpheratz

Aquarius
The Water-Bearer

The entire region in which Aquarius lies has watery connections—the constellation is surrounded by the Fishes, the Sea Goat, the Southern Fish, and the Whale. The Babylonians called this region "the Sea," with all constellations being under the control of Aquarius. The Egyptians believed that Aquarius caused the annual flooding of the Nile River, and so it was a very important constellation to them. It is no coincidence that the hieroglyph for running water is now the astrological sign for Aquarius.

Traditional star maps depict Aquarius as a youth pouring water from a pitcher. Beneath his feet, the gushing water ends up in the mouth of Piscis Austrinus, the Southern Fish. The beautiful youth was probably Ganymede, son of the king of Tros, founder of the city of Troy. Zeus wanted Ganymede as his favorite and sent an eagle to carry him off to Mount Olympus. There he became the cupbearer of the gods, delighting all with his beauty.

Starry triangle

Aquarius is not an easy constellation to make out. It can probably best be located by reference to the Square of Pegasus to the north. Even its lead star Alpha (α) is disappointingly dim. There are a few highlights, however; one is M2, which makes a triangle with the stars Alpha and Beta (β).

It is one of the brightest globular clusters in the sky and just on the limit of naked-eye visibility. Binoculars and small telescopes show it well.

What the astrologers say

Aquarius is a constellation of the zodiac lying between Pisces and Capricornus. Currently, the Sun passes through Aquarius from February 16 to March 11. In astrology, Aquarius is the eleventh sign of the zodiac, covering the period January 20 to February 18. Astrologers believe that Aquarians are cool and detached, even aloof; they tend to conceal emotions and find it difficult to fall in love or relate to partners.

Larger telescopes are needed to spot the Helix nebula, near the southern end of the constellation. It is one of the nearest planetary nebulae and the Hubble Space Telescope has discovered some incredible "cometary knots" in it.

Alpha and Beta mark the shoulders of the figure, who the ancient Greeks associated with the handsome Ganymede.

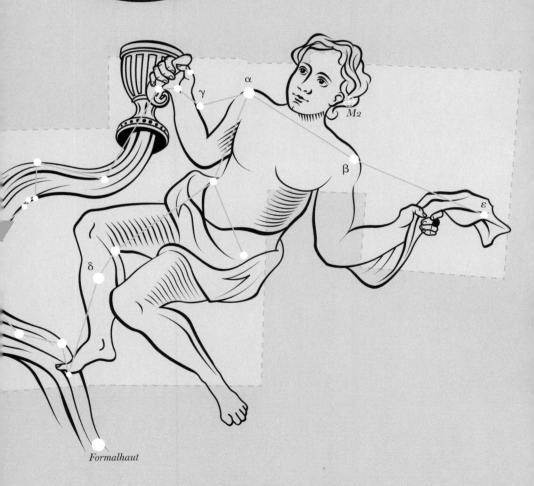

Formalhaut

Aries

The Ram

Aries is best located by following a line from the unmistakable cluster of the Pleiades to the Square of Pegasus. It is located about halfway between the two.

In mythology, the ram was the source of the Golden Fleece that Jason and the Argonauts went to steal (see pages 48–49). The ram was a magical creature that could speak, think, and fly through the air. Hermes gave the ram to the two children of King Athamas, Helle and her brother Phrixus, who fled on it to escape from their hated stepmother. Helle unfortunately fell off into a strait that became known as the Hellespont. Phrixus safely reached Colchis on the Black Sea and sacrificed the ram to show his gratitude for being saved. He gave the Golden Fleece to Aeetes, king of Colchis, who set the ferocious dragon Draco to guard it.

What the astrologers say

Aries is a constellation of the zodiac, lying between Pisces and Taurus.

Currently, the Sun passes through Aries between April 18 and May 21. In astrology, Aries is the first sign of the zodiac, covering the period from March 21 to April 19, for that is what it did in classical times. March 21 is around the time of the vernal, or spring, equinox, when the Sun crosses the celestial equator traveling north. On this date, the path of the Sun through the heavens (the ecliptic) and the celestial equator intersect. This point is known as the First Point of Aries. However, because of precession (a wobbling of the Earth's axis), the Sun is no longer in Aries at this time, but in Pisces (see pages 84–85).

Astrologers tell us that Arians are energetic, impulsive, and extroverted. They are believed to be courageous and adventurous, and natural-born leaders. Warm and passionate. In relationships, they frequently find the chase more exciting than the surrender.

The three brightest stars—Alpha, Beta, and Gamma—outline the head of the ram with the golden fleece.

The two main stars of Aries are of the second magnitude, but it is the fainter Gamma that is the more noteworthy. In small telescopes, Gamma shows up as a lovely double, with the pair of stars of equal blue-white brilliance.

Auriga
The Charioteer

This familiar northern constellation depicts the driver of a horse-drawn chariot, and originally included the chariot as well. Auriga is usually identified with Erichthonius, who became king of Athens. His father was the Greek god of fire, Hephaestus (Vulcan in Roman mythology), but he had no mother.

According to the myth, this came about because Hephaestus had lusted after the conspicuously virginal goddess Athene, and one day attempted to seduce her. She fought him off and escaped, while he scattered his seed on the earth. Shortly afterward, the earth gave birth to a boy-child, Erichthonius. Then, by chance, Athene found him and brought him up, teaching him the skills of horsemanship. Erichthonius then developed the four-horse chariot. Later he became king of Athens and established the cult of Athene worship there in gratitude.

Auriga is usually depicted with a goat draped over his left shoulder. The brightest star in the constellation, Capella marks its position—the name means she-goat. The goat is generally identified with Amalthea, the goat that suckled Zeus as a baby. The goat's kids are also present, held by Auriga's left arm. They are marked by a triangle of fainter stars and are known as the Kids, or the Heidi.

The Goat Star

Auriga is not difficult to spot because of its lead star Capella. It forms a compact triangle of constellations with Gemini and Taurus. Capella is often referred to as the Goat Star. It is the sixth brightest star in the sky and it emanates a yellowish light similar to that given off by our own Sun. But there the resemblance ends. Two of the three stars that mark the Kids, Epsilon (ε), and Zeta (ζ) are also binary, or double-star systems. They are eclipsing binaries, whose brightness periodically dims when one of the stars eclipses the other. Since Auriga straddles the Milky Way, it is a good subject for sweeping with binoculars. Several bright open clusters at the edge of the Milky Way, near the star Theta (ϑ) are visible.

Auriga's lead star, Capella, is nearly 100 times brighter than the Sun, a yellow dwarf less than one-tenth the size of Capella. Capella is a spectroscopic double—the stars are so close together that they can be detected only in a spectroscope.

No legends explain why the charioteer has a goat on his arm.

δ

Menkalinan

β

α   Capella

ε

η   ζ

θ

ι

β Tauri

Boötes
The Herdsman

This prominent northern constellation has an unmistakable shape of a traditional kite. It is easily located by following the curve of the handle of the Big Dipper (the Plow), the prominent part of the Great Bear (Ursa Major). Interestingly, bears also feature in the story of this constellation. It was sometimes known as Arctophylax, meaning the Bear Keeper. And Boötes was sometimes referred to as the Bear Driver, rather than the Herdsman, chasing the Great Bear and the Little Bear (Ursa Minor) across the sky. Boötes holds the leads of the hunting dogs, represented by the constellation Canes Venatici, that pursue the bears across the sky.

Boötes is usually thought to represent Arcas, who was an offspring of Zeus and the nymph Callisto. Callisto had earlier taken a vow of chastity and became a companion of Artemis. However, cunning Zeus appeared to her in the guise of Artemis, and when she discovered his real identity, it was too late. So she gave birth to Arcas. Artemis was furious with Callisto, and Zeus changed her into a bear to help her escape Artemis's rage.

One day, when he was a man, Arcas was out hunting and came across Callisto in the guise of a bear. She recognized him, but he did not know his mother. He gave chase and she sheltered in a sacred place in which discovery would mean death. To protect her and Arcas, Zeus placed them in the heavens. There are several variations of this story, and indeed other stories about the origin of the constellation.

The bear's tail

Arcturus translated means bear's tail, and it is found at the tail end of the constellation's kite shape. Of first magnitude, it is a red giant, nearing the end of its life, and to the eye has a definite reddish tinge.

There is little else of naked-eye interest in this constellation, but it has some fine double stars visible through small telescopes. The best must be Epsilon (ε), seen as a beautiful yellowish-orange and bluish-green pair. Xi (ξ) and Mu (μ) are also colorful pairs.

Boötes' leading star, Arcturus, is the brightest in the Northern Hemisphere and the fourth brightest in the whole sky. About twenty-five times the diameter of the Sun, Arcturus is also 100 times brighter.

One story tells how Boötes gained his place in the heavens for inventing the ox-drawn plow.

θ

λ

μ

β

γ

δ

ρ

ε

α

ξ

Arcturus

υ

Cancer
The Crab

Cancer is one of the faintest of the constellations but can be found quite easily east of a line connecting the bright stars in Gemini—Castor and Pollux—with Procyon in Canis Minor to the south.

In mythology, Cancer represents a crab that became involved in battling with Hercules while he was fighting the dreaded hydra. Hercules was the son of Zeus and the nymph Alcmene, the product of one of the king of the gods' many brief illicit affairs. Zeus's wife, Hera, thereafter hated Hercules and did her best to destroy him, first making him kill his wife and children in a fit of madness, and then getting him involved in virtually impossible tasks, or labors, which would surely kill him.

Having defeated the Nemean lion on his first labor, Hercules took on the multiheaded hydra and vanquished this monster as well. Hera then sent a crab to attack Hercules, but to little avail. He crushed the crab underfoot, whereupon Hera placed it in the heavens.

Bees around the hive

The most interesting feature of this faint constellation is centered on the middle star, Delta (δ). If you look at Delta on a really dark night, you can see a swarm of stars close by. This swarm is named Praesepe, and is called the Beehive because the stars look rather like bees swarming around the hive.

What the astrologers say

Cancer is one of the constellations of the zodiac, lying between Gemini and Leo. Currently, the Sun passes through Cancer between July 20 and August 10. In astrology, Cancer is the fourth sign of the zodiac, covering the period from June 22 to July 22, for that is what it did in classical times. June 22 is the time of the summer solstice, when the Sun reaches its farthest point north of the Equator. This extremity defines the northernmost point of the Tropics, and is called the Tropic of Cancer. However, because of precession, the Sun is now in Gemini at the summer solstice, so in theory it should now be renamed the Tropic of Gemini.

Astrologers say that Cancerians tend to be emotional people, quiet and even sullen at times. But they are also caring, kindly people; home-loving and dependable, they make good partners, but can become vindictive if they are crossed in love.

Near Cancer's middle star, Delta, lies an open cluster of stars called Praesepe. It looks better viewed with binoculars and best using a small telescope. In all, Praesepe probably contains at least 300 individual stars.

The Italian astronomer Galileo first resolved the Praesepe cluster into individual stars.

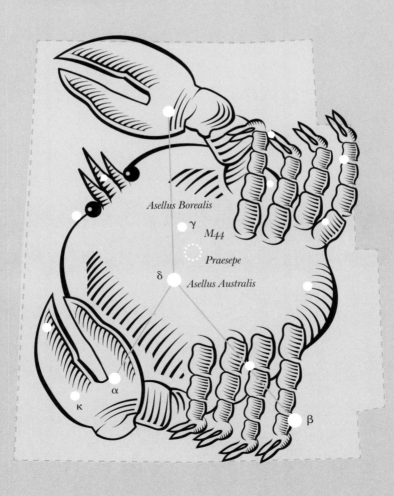

Asellus Borealis

γ

M44

Praesepe

δ Asellus Australis

α

κ

β

Canis Major
The Great Dog

Canis Major represents one of the dogs that accompanied the hunter Orion. In the heavens it spends its time chasing the hare (the constellation Lepus) at Orion's feet. Orion's other dog, Canis Minor (the Little Dog), is located across the Milky Way from Orion and Canis Major. In mythology, this constellation is associated with the dog Laelaps, which could outrun anything it chased. Eventually, though, it went in pursuit of a fox that ran so fast that nothing could catch it, and the two seemed destined to race each other forever. But Zeus stepped in and turned them both to stone, placing Laelaps in the heavens as Canis Major.

The brightest star of all

The constellation is dominated by its lead star Sirius, which is the brightest star in the heavens. Unsurprisingly, it is popularly known as the Dog Star, and has possessed this name since classical times. The name Sirius means scorching. The Greeks thought that it was responsible for the heat of summer, since in July and August it rose in the dawn sky just before the Sun. These hot, sultry days were named the Dog Days as a result.

The early Egyptians did not recognize any dog constellation, but they regarded Sirius as vitally important. They called it Sothis, or the Nile Star. They even worshiped it because it happened to appear in the dawn sky just before the Nile flooded each year—they depended on this flooding to irrigate and nourish their land for agriculture.

Sirius has a magnitude of around -1.5 (negative magnitudes indicate exceptionally bright stars). It is about twice as bright as the next brightest star, Canopus. Strangely, both these stars lie relatively close together in the heavens, though this is just a coincidence.

Sirius is not a single star, but a binary. The other star is a tiny white dwarf. Indeed it was the first star of this type to be discovered, in 1862. It is named the Companion of Sirius, with the popular nickname of the Pup.

Sirius, Canis Major's lead star, appears so bright to us because it lies relatively close—a little under nine light-years away. Sirius is only about twenty-five times brighter than our Sun, while Rigel, lying more than 100 times further away in Orion, is some 60,000 times brighter.

In ancient Egypt, this constellation was depicted as a cow lying down, with Sothis (Sirius) between its horns.

Capricornus
The Sea Goat

Capricornus is a relatively faint constellation, but it is one of the most ancient and is always associated with water and the oceans. In Babylonian times, it was known as the Goat-Fish and was considered to rule the surrounding heavens, from which the rivers Tigris and Euphrates flowed.

The Greeks also saw Capricornus as a strange creature that had the head and forelimbs of a goat but the tail of a fish. They identified it with the pipe-playing god Pan, who was also a strange hybrid creature with goat's legs and horns. Pan was the great god of the forests and meadows, through which he forever wandered, playing and dancing with nymphs. One day he was pursuing the nymph Syrinx with amorous intent, but her sisters turned her into reeds as he was about to pounce. As he sighed, his breath blew over the reeds, which gave off musical sounds. He cut off a number of them of different length and bound them together to form the pipes of Pan, which are also called the syrinx.

What the astrologers say

Astrologers always call this constellation Capricorn. It is the tenth sign of the zodiac, covering the period from December 22 to January 19, as in classical times. Currently, the Sun passes through Capricornus between January 19 and February 16.

December 22 is near the time of the winter solstice, when the Sun reaches its farthest point south of the Equator. This extremity defines the southernmost point of the Tropics, and is called the Tropic of Capricorn. However, because of precession, the Sun is now in the constellation of Sagittarius at the winter solstice, so we should now really refer to the Tropic of Sagittarius.

According to astrologers, Capricorns tend to be hard-working and reliable. They may have a hard exterior, but are often sensitive underneath. They crave respect and recognition. Capricorns are loyal but may sometimes seem rather cool partners, reluctant to show their true feelings.

Capricornus is an inconspicuous constellation with only a few third-magnitude stars. It is the smallest constellation of the zodiac, lying in a "watery" region of the heavens between Sagittarius and Aquarius.

The keen-sighted can spot that Alpha is double, but in reality its two stars are hundreds of light-years apart.

Deneb Algedi

Dabih

Carina
The Keel

This is one of the stunning far southern constellations beyond the gaze of most Northern Hemisphere astronomers. The ancients did not recognize Carina as a separate constellation, but as part of the much larger constellation Argo Navis, the Ship of the Argonauts. Carina represents the keel of this craft, while Vela represents the sails and Puppis the poop (or stern). The French astronomer Nicolas Lacaille split the constellation into three parts in 1763.

The Argonauts feature in one of the most famous legendary tales of Ancient Greece—Jason's quest for the Golden Fleece. The Golden Fleece was found at Colchis, in a cave guarded by the dragon Draco. Jason took with him fifty great heroes, including Castor and Pollux (the twins in the constellation Gemini, see pages 66–67) and the musician Orpheus.

On their journey to and from Colchis, Jason and his crew had many adventures. For example, they had to pass through the Clashing Rocks, which opened and shut like sliding doors and would crush anything caught between them. Jason got through safely by sending a dove ahead to make the Rocks shut, then rowed for dear life when they briefly opened up again. At Colchis, Jason fell in love with Medea, daughter of King Aeetes, who possessed the Golden Fleece. She was a sorceress and helped the Argonauts steal it and returned with them to Greece. Jason left the ship *Argo* in a grove sacred to Poseidon at Corinth.

Sailing the Milky Way

Carina and the rest of Argo Navis are set in one of the most dazzling regions of the Milky Way. It adjoins Crux, the Southern Cross, and Centaurus, the Centaur. In the region of the Milky Way, halfway between the False Cross and the Southern Cross, is one of the most brilliant nebulae in the heavens, the Eta Carinae Nebula NGC3372. In the 1800s, the star Eta (η) Carinae itself flared up to become brighter than Canopus. It is one of the largest of the known stars and is highly unstable, constantly puffing out vast clouds of gas and dust.

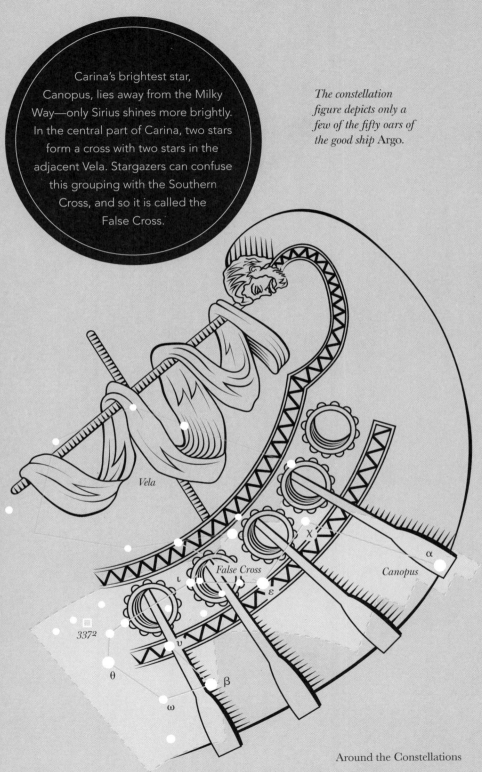

Carina's brightest star, Canopus, lies away from the Milky Way—only Sirius shines more brightly. In the central part of Carina, two stars form a cross with two stars in the adjacent Vela. Stargazers can confuse this grouping with the Southern Cross, and so it is called the False Cross.

The constellation figure depicts only a few of the fifty oars of the good ship Argo.

Vela

χ

False Cross

ι

ε

α

Canopus

υ

337²

θ

β

ω

Cassiopeia
The Vain Queen

Cassiopeia lies in the northern heavens between Cepheus and Andromeda. In legend, Cepheus was the king of Ethiopia, Cassiopeia was his queen, and Andromeda was their daughter. Legendary Ethiopia was not the Ethiopia we know today, but a region near present-day Jordan and Egypt.

Queen Cassiopeia was very attractive, but also vain. She liked doing nothing better than sitting in a chair in front of a mirror and combing her long hair. One day she thought she looked particularly glamorous, and boasted that she had to be more beautiful even than the lovely sea nymphs known as the Nereids.

Learning of this, the sea nymphs were understandably upset. There were some fifty Nereids in all and one, named Amphitrite, was married to the god of the sea, Poseidon. She asked her god husband to punish Cassiopeia for her vain boast, which he did. He sent a terrible sea monster called Cetus (see pages 54–55) to lay waste the kingdom of Cepheus and Cassiopeia. To put an end to the carnage, Cepheus and Cassiopeia had to offer their daughter Andromeda as a sacrifice to the monster, and she was duly chained to some rocks to await her fate. But, in the nick of time, the hero Perseus came along and slew the monster (see Andromeda, page 32–33).

Cassiopeia's punishment wasn't over. On her death, she was placed in the sky as a constellation close to the Pole Star. There, she sits in her chair, still combing her hair, and condemned to circle for all eternity, often hanging upside-down.

Circling for eternity

Cassiopeia has one of the most recognizable shapes among the constellations, a distinct W, formed by its five main stars. From much of the Northern Hemisphere, Cassiopeia is circumpolar—it always stays above the horizon. This is why Queen Cassiopeia is condemned to circle for all eternity.

The vain Queen Cassiopeia is one of the most undignified of postures in the heavens.

ε

δ

γ

Ruchbah

υ

β

Shedir   α

Centaurus
The Centaur

This constellation gets its name from the mythical half-man, half-horse creatures that lived in northern Greece. The majority were a wild and lawless bunch, that liked nothing better than feasting, drinking, and fighting. Eventually, they reaped the just rewards for their riotous lifestyle. One day, at a wedding feast for the king of the Lapiths, a drunken centaur attacked the bride. A great battle ensued in which the Lapiths defeated the centaurs and drove them away. This battle is the subject of some magnificent sculptures in the temple of Zeus at Olympia.

The centaur in the constellation was the offspring of an illicit liaison between Cronos, king of the Titans, and a sea nymph named Philyra. The king's wife, Rhea, caught them in the act, so Cronos turned himself into a horse and galloped away. Later, Philyra bore their half-man, half-horse son, Chiron.

Chiron had none of the wildness of the other centaurs. He became wise and learned, and proved to be a brilliant teacher of the arts and of healing. One of his notable pupils was Apollo's son Asclepius, who grew up to be the Greek god of medicine. Chiron was killed accidentally by Hercules, who winged him with a poisoned arrow. The arrow was tipped with the Hydra's blood, which was lethal.

The Southern Pointers

Centaurus is a southern constellation, in which the Centaur's feet straddle the unmistakable Southern Cross, the constellation Crux (see pages 56–57). Its two brightest stars, Alpha (α) and Beta (β) Centauri, are two of the brightest stars in the sky. They make good pointers to the Southern Cross.

Alpha Centauri is the closest bright star to us, being at a distance of about 4.3 light-years. Near it is a faint red dwarf star named Proxima Centauri, which is actually the nearest star.

Much of Centaurus lies in the Milky Way, so it is rich in dense star clouds, nebulae, and clusters. Omega (ω) Centauri is the finest example in the whole heavens of a globular cluster— a globe-shaped mass of stars packed tightly together. Easily visible to the naked eye, Omega Centauri looks magnificent through binoculars.

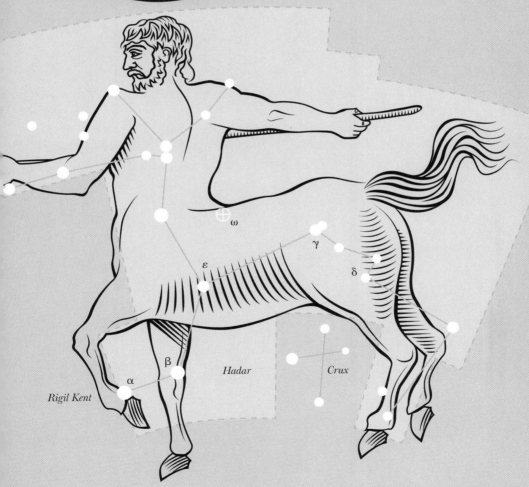

Follow a line with your eyes from Beta Centauri through Epsilon (ε) to an equal distance beyond, where there is a bright patch that looks like a blurry star. It is not just one star, however, but a grouping of millions of stars called Omega (ω) Centauri.

As depicted, the Centaur is about to kill a wolf (the adjacent constellation Lupus) as a sacrifice.

ω

γ

ε

δ

β

α

Rigil Kent

Hadar

Crux

Cetus

The Sea Monster or Whale

This constellation is the fourth largest in the heavens and it is an ancient one—Ptolemy listed it as having twenty-two stars nearly two millennia ago. Cetus is part of the ocean-related group that spans a vast expanse of the heavens, which also includes nearby Pisces (the Fishes, see pages 84–85), Aquarius (the Water-Bearer, see pages 34–35), and Piscis Austrinus (the Southern Fish).

Cetus has always been depicted as a bizarre-looking monster of tremendous size and horrific appearance. Its enormous head has gaping jaws studded with vicious teeth, and its stubby forelimbs end with sharp claws. Its long body is covered with scales, and its tail is coiled like that of a sea serpent. Although Cetus is often identified with the Whale, there is nothing whale-like about this creature, apart from its huge size. Ancient peoples would have been familiar with whale skeletons even if they never saw a whale in the flesh, so they undoubtedly clothed such skeletons in their imagination and came up with the monster.

Cetus "swims" in the heavens not so far from Andromeda, representing the chained maiden who in mythology was presented as a sacrifice for it. The sacrifice had been made because her boastful mother had offended the gods (see pages 32–33). Before Cetus could get at Andromeda, he was killed by the hero Perseus (see page 50). One story says it was put to the sword by Perseus, another that it was turned to stone when Perseus exposed it to the eyes of Medusa, whose head he was carrying.

Mira the Wondrous

Cetus is a sprawling constellation, and a faint one. Its two brightest stars Alpha (α) and Beta (β) are about third magnitude, set respectively in the head and tail of the monster.

About one-third of the way along the line from Alpha to Beta is the most interesting star for observers. It is designated Omicron (o) in the constellation, but astronomers also call it Mira, meaning the Wonderful One. Mira Ceti is wonderful because it varies in brightness markedly over time. At its brightest it is third magnitude, like Alpha and Beta, and is easily visible to the naked eye. But then it gradually fades to about magnitude 10, which renders it invisible to the naked eye—even with ordinary binoculars. Later it gradually regains its former brightness.

The sea monster Cetus set out to devour Andromeda, but was thwarted by the hero Perseus.

Cetus' most interesting star, Omicron—or Mira Ceti—is the prototype of the long-period variable stars, which are known as Mira variables. It is a huge red giant, which varies in brightness over a period of about eleven months.

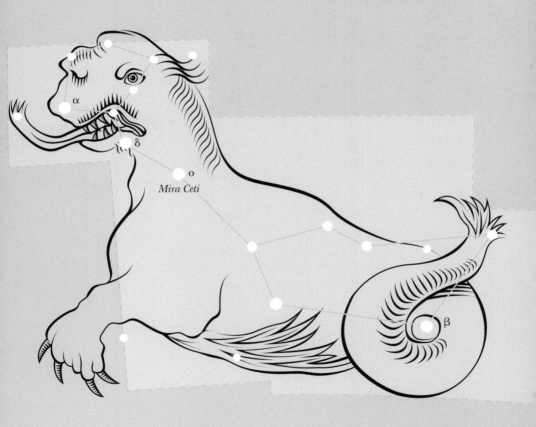

α

δ

ο

Mira Ceti

β

Crux

The Southern Cross

In the Southern Hemisphere, no constellation is more famous than the Southern Cross. It lies so far south that astronomers in northern Europe and the upper reaches of North America can never see it. The Ancient Babylonians and the Greeks were far enough south to see it. In those days, it was regarded as being part of the constellation Centaurus, located as it is between the Centaur's legs.

Crux only came to be recognized as a separate constellation some time in the sixteenth century. That is also when European explorers first began sailing the southern oceans and using the Southern Cross for navigation. The cross was a boon for navigators because its long axis points almost exactly to the south celestial pole. It was as important to sailors in the Southern Hemisphere as the North Star (Pole Star) was to sailors in the Northern Hemisphere.

Crux can be confused with the False Cross between nearby constellations of Carina and Vela. The best way to locate it is by using the bright Centaurus stars Alpha and Beta as pointers.

The smallest of them all

Small it may be, but Crux contains much of interest. Of its four main stars, two are first magnitude, one is second, and the other is third.

Close to Alpha (α), also called Acrux, there appears to be a dark hole in the Milky Way. It is actually a dark, dense nebula of gas and dust that obscures the stars behind it and is aptly named the Coal Sack.

Close to Beta (β) and bordering on the Coal Sack, is a group of colored stars clustered around Kappa (k). They are easily visible to the naked eye and stunning when viewed through binoculars. Astronomer John Herschel, son of William who discovered Uranus, named this open cluster of stars the Jewel Box. And it does seem to sparkle like sapphires, rubies, and diamonds.

Crux is the smallest of all the constellations. It is about one-twentieth the size of the largest constellation, Hydra. Because it is embedded in a brilliant part of the Milky Way, it can be difficult to find right away.

Small but spectacular, the Southern Cross sits beneath the hind legs of Centaurus, the Centaur.

CENTAURUS

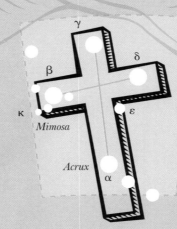

β

γ

δ

κ

ε

Mimosa

Beta Centauri

Acrux

α

Alpha
Centauri

Cygnus
The Swan

Cygnus is one of the few constellations that looks passably like the figure it is supposed to represent. With only a little imagination one can flesh out its pattern of bright stars into a swan, with wings extended in flight and long neck outstretched.

To the Greeks, Cygnus represented the king of the gods, Zeus, in disguise. Among the many women he lusted after was Leda, the queen of Sparta. He turned himself into a swan and slept with her on the riverbank. Later that night, Leda also slept with her husband, the king. Leda became pregnant, but instead of giving birth to babies, she produced two enormous eggs.

From one egg hatched Castor and Pollux (the twins in Gemini, see pages 66–67); from the other Clytemnestra and Helen. Castor and Clytemnestra were fathered by the Spartan king, Pollux and Helen by Zeus.

Helen grew to be the most beautiful woman in the world, the famed Helen of Troy, with "the face that launched a thousand ships and sank the topless towers of Illium." This refers to the fact that she was the cause of the Trojan Wars between Sparta and Troy.

The Swan's tail

Astronomically, Cygnus looks beautiful to the naked eye, flying along the Milky Way. Its brightest star, Deneb, in the Swan's tail, is first magnitude. It lies much farther away than most other bright stars. But we still see it shining brilliantly because it is exceptionally luminous, shining with the power of more than 60,000 suns. Deneb is one of the three bright stars that make up the celebrated Summer Triangle, along with Vega in Lyra and Altair in Aquila.

The Milky Way is worth sweeping in binoculars. The region between Deneb and the fainter Xi (ζ) reveals a large nebula. Telescopes are needed to bring out its uncannily life-like shape of North America, and it is indeed called the North American Nebula.

The beak of the Swan is marked by the third magnitude star Albireo. This star is a favorite for small telescope observers. It is a beautiful double, with blue and yellow components.

It is not surprising that an alternative name for this constellation is the Northern Cross.

α

Deneb

ξ

δ

γ

ε

β

Albireo

Dorado
The Swordfish

Dorado is one of the more modern constellations, first depicted on star maps in the late sixteenth century. It lies in a fairly undistinguished part of the far southern heavens, not far from the south celestial pole.

The constellations surrounding Dorado—Mensa (the Table), Pictor (the Painter), Caelum (the Chisel), Reticulum (the Net), and Hydrus (the Little Snake)—are also recent in origin. There is no classical mythology associated with these constellations, created out of what are called amorphae, or the "left-over" stars.

Star island in space

The line of fairly faint stars that marks out Dorado can be found by locating the beacon star Canopus and moving in the direction of the first magnitude Achernar in Eridanus. Slightly to the south of this line you will see Dorado's main claim to fame; a blurry patch, easily visible to the naked eye.

This blurry patch looks like a rather detached section of the Milky Way, but it isn't. Binoculars cannot resolve it into

stars like they can the Milky Way. But small telescopes will begin to spot in it bright stars, and large telescopes will pick up masses of fainter stars. It seems to be a star system much farther away than the Milky Way. And so it is. It is another star island in space—another galaxy, and the next nearest galaxy to our own.

Cloud-like to the naked eye, it is known as the Large Magellanic Cloud (LMC). Very much smaller than our own galaxy and irregularly shaped, the LMC is named for the Portuguese navigator Ferdinand Magellan, who would have been one of the first Europeans to see it. He commanded the first expedition to sail around the world, which set out in 1519. However, Magellan never completed this first circumnavigation as he was killed in the Philippines two years later.

And why is it known as the "Large" Magellanic Cloud? Well, following a line from Canopus through the LMC, you come to a smaller fuzzy patch. This is the Small Magellanic Cloud, another slightly more distant galactic neighbor.

The Large Magellanic Cloud in Dorado contains much the same mix of ordinary stars, variables, clusters, and nebulae as in our galaxy. One of these nebulae can be seen with the naked eye. It is called Tarantula Nebula, for its resemblance to the notorious spider.

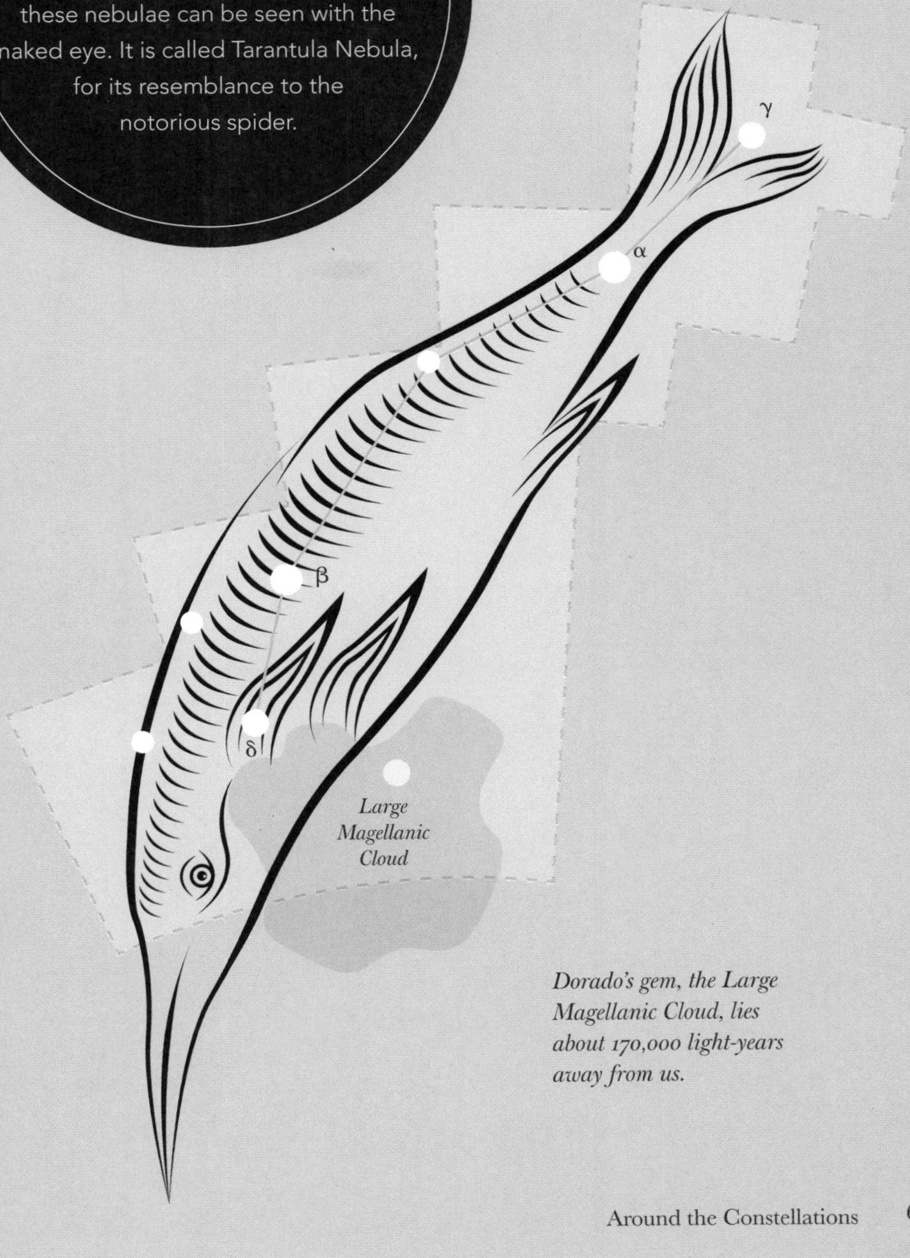

nopus

γ

α

β

δ

Large
Magellanic
Cloud

*Dorado's gem, the Large
Magellanic Cloud, lies
about 170,000 light-years
away from us.*

Draco
The Dragon

Draco winds itself, serpent-like, nearly halfway around the celestial north pole, marked by Polaris, the Pole Star. It is so far north that, from most of the Northern Hemisphere, it is circumpolar; it never sets and is always visible over the horizon.

In mythology, the dragon that never sets became the dragon that never sleeps. Known as Ladon, this monster had 100 heads and was an eternally vigilant guardian of the sacred golden apples in the exquisite Garden of the Hesperides, located in the most remote western limits of the world. The Hesperides were the daughters of Atlas and Hesperus, the evening star (Venus). They had been entrusted with the task of guarding the apples, but could not resist picking and eating them.

As the eleventh of his labors, the most heroic of the Greek heroes, Hercules, set out to steal the golden apples for his cousin, Eurystheus, king of Argos. In seeking the garden, he had many brushes with death having to overcome many attackers, including a lion and a terrifying eagle. Eventually he arrived at the Garden of the Hesperides. There he slew Ladon and made off with the apples. One version of the story says that Atlas, who carried the world on his shoulders, helped him. Hercules persuaded Atlas to pick the apples while he temporarily carried the world on his shoulders.

Thuban and Pole Stars

Draco winds its way around Polaris. But Polaris has not always been the Pole Star. As the Earth spins around in space, it wobbles, rather like a spinning top. And this wobbling motion, known by astronomers as precession, makes the Earth's axis point in different directions in space at different times.

The axis points to Polaris, so this is now the Pole Star. Four thousand years ago, in ancient Egypt, the axis pointed to the star Alpha (α) in Draco, called Thuban (meaning serpent's head). This will again become the Pole Star in about 22,000 years, for it takes the Earth some 26,000 years to go though each cycle of precession.

Draco is a sprawling serpent-like constellation—the eighth largest in the heavens. The Great Pyramid of Giza, built around the time when Thuban was the Pole Star, has a passage lined up with it.

In this sprawling constellation, Gamma is the brightest star, not Alpha.

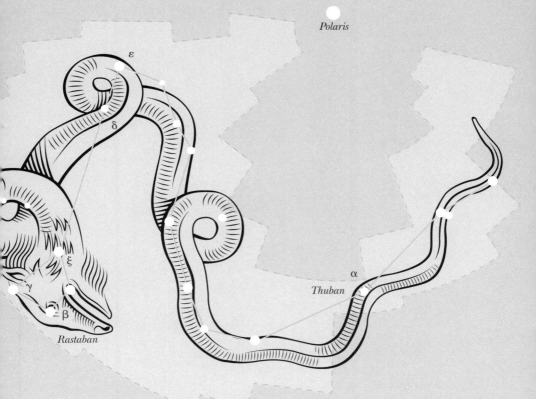

Polaris

ε

δ

ξ

γ

ε β

Rastaban

α

Thuban

Eridanus
The Winding River

Meandering its way from Orion near the celestial equator and deep into the southern celestial hemisphere, Eridanus is the longest of all the constellations. It has been associated over time with different rivers, real and imaginary. It was one of the rivers that flowed into the great river ocean, Oceanus, which girdled the universe. To the Babylonians, it was the heavenly representation of the Euphrates, one of the two rivers in Mesopotamia (which means between two rivers—the other being the Tigris). To the Egyptians the river was the Nile.

The mythical Eridanus River figures in the terrifying adventures of Phaeton, one of the sons of the god of light, Helios (the Greek word for the Sun). One day, to prove to doubters that he was of divine origin, he begged his father to let him drive the Sun's chariot across the sky for one day.

Eventually, Helios agreed and handed over the reins. The impetuous horses immediately noticed the inexperienced hands, and set off at a mad pace. They galloped wildly through space and went perilously close to Earth—so close that the soil began to burn and the rivers

began to dry up. The whole universe indeed was in danger of catching fire. To avert complete disaster, Zeus struck Phaeton with a thunderbolt, and the hapless youth tumbled back to Earth and plunged into the Eridanus River, where he drowned. His sisters, the Heliads, came to mourn him and their tears turned to amber, which washed up on the riverbank.

The end of the river

Most of the stars in Eridanus are faint, occupying a rather empty part of the heavens that also includes the other sprawling constellation of faint stars, Cetus (see pages 54–55). Only Achernar at the southern extremity of Eridanus is bright.

Astronomically, one of the most interesting stars in the constellation is Eta (η). Eta Eridani happens to be a star very much like the Sun. At a distance of about eleven light-years, it is one of the closest of the Sun-like stars. It might even have a planetary system around it, which has made it a favorite target for SETI researchers. (SETI stands for search for extraterrestrial intelligence).

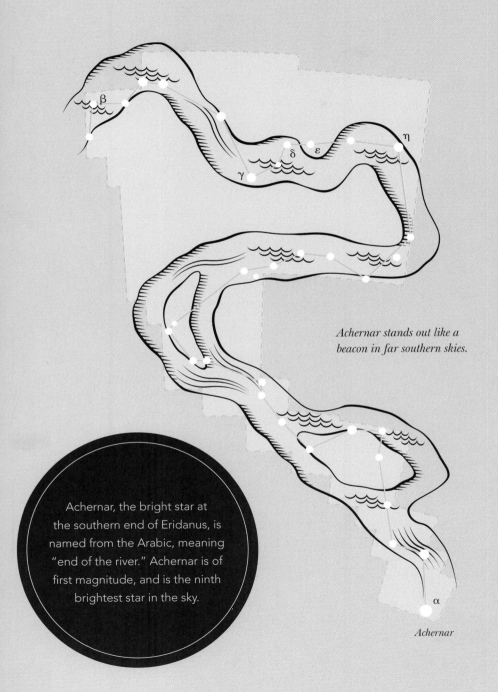

β

η

δ ε

γ

Achernar stands out like a beacon in far southern skies.

Achernar, the bright star at the southern end of Eridanus, is named from the Arabic, meaning "end of the river." Achernar is first magnitude, and is the ninth brightest star in the sky.

α

Achernar

Gemini
The Twins

Gemini is one of the outstanding constellations of the winter skies in the Northern Hemisphere, dominated by its brilliant twin stars Castor and Pollux.

It puts on a magnificent light show along with Auriga, Taurus, and the stunning Orion. Castor and Pollux form part of a great oval of first magnitude stars that also include Procyon, Sirius, Rigel, Aldebaran, and Capella.

In mythology, Castor and Pollux were the sons of the fair Leda. Zeus seduced her in the guise of a swan, which was placed in the heavens as Cygnus (see pages 58–59). As a result of this union, Leda laid two eggs, which hatched into four children; two of these children were Castor and Pollux. These two infants were identical twins, they grew up together and became inseparable. But they had different talents: Castor became an incomparable horseman, while Pollux was a formidable boxer. Among other adventures, they sailed with the Argonauts and calmed the seas that threatened to sink their ship. As a result the sea god Poseidon made them the protectors of sailors.

The magnificent duo

Both Castor and Pollux are first magnitude stars. Pollux is the brighter of the two and is a richer golden color. Castor, however, is more interesting because it is a multiple star. Small telescopes will reveal two or three components, while larger instruments reveal that each is a binary, two-star system. This makes Castor a six-star system. The Twins have their feet in the Milky Way, which is rich in star clouds and clusters.

What the astrologers say

Gemini is a constellation of the zodiac, lying between Taurus and Cancer. Currently, the Sun passes through Gemini between June 21 and July 20. In astrology, Gemini is the third sign of the zodiac, covering the period from May 21 to June 21. Geminis are said to be lively and communicative characters, and are clever in how they express themselves. However, they can be moody. Their talents suit them for work as journalists and sales executives.

Among Gemini's stars, William Herschel discovered Uranus in 1781, after first thinking that it was a comet, and Clyde Tombaugh discovered Pluto in 1930.

Gemini is one of the brightest constellations in the zodiac, with Castor and Pollux outstanding.

α

Castor

β

Pollux

υ

M35

δ

ε

η

μ

ν

Alhena
γ

Hercules

The Great Hero

The northern constellation Hercules is visually not as impressive as it perhaps should be given that it represents the most famous and indefatigable of all Greek heroes. We know him best by his Latin name, but the Greeks called him Heracles. This was probably because he owed his prowess to Zeus' wife Hera. It was she who suckled him, but he was not her son. He was the product of one of Zeus's dalliances, with the beautiful Alcmene. When Hera learned the truth of Hercules' parentage, she hated him from that day forth.

After Hercules had married, Hera made him go mad, whereupon he murdered his wife and children. Hercules wanted to atone for his grievous crime and so he consulted the Oracle at Delphi, who commanded that he spend twelve years serving his cousin, Eurystheus. During this time Eurystheus set him twelve tasks, known as the Labors of Hercules.

Among his most celebrated labors, Hercules killed the Nemean lion (represented in the sky by Leo, see pages 72–73) by strangulation, since arrows would not pierce its skin— Hercules used its skin to make a cloak. He then vanquished the multiheaded Lernean hydra that breathed poison and dipped his arrows in the hydra's blood, which made them deadly. In one day he cleaned the Augean Stables, which had the accumulated dung of decades. He killed the dragon, Ladon that guarded the golden apples of the Garden of the Hesperides. Then, in his final labor, he descended to the Underworld and killed Cerberus, the guardian of the gates of hell (Hades).

After many other exploits, Hercules ironically fell victim to the hydra's poison in a tragic chain of events. In perpetual agony, Hercules took his own life on a funeral pyre.

The super cluster

Astronomically, the highlight in Hercules is M13, an outstanding globular cluster. It looks like a fuzzy blob in binoculars and is magnificent in small telescopes, which begin to resolve the individual stars.

M13

ζ

ε

δ

β

γ

α

The globular cluster in Hercules called M13 is made up of hundreds of thousands of stars. It is the finest cluster in the northern skies, and is visible to the naked eye.

In the heavens, Hercules wields a club. His posture is not unlike that of another celebrated Greek, Orion.

Hydra
The Water Snake

Out of all eighty-eight constellations, Hydra is the largest. It winds its way, serpent-like, more than a quarter of the way around the celestial sphere. It runs roughly parallel to the celestial equator, south of the three constellations of the zodiac, Cancer, Leo, and Virgo.

In mythology, the hydra was a nine-headed serpent, born of the monster Typhon. The middle head was immortal. It inhabited the marshes near Lerna, in the Peloponnese, ravaging and terrorizing the region. It had deadly poisonous breath so those who felt it died an agonizing death.

The hero Hercules was set the task of killing the hydra as the second of his twelve labors. Accompanied by his chariot-driver Iolaus, Hercules arrived at Lerna and found the monster at the spring of Amymone. By firing flaming arrows into the marshes, he forced the hydra out into the open. He sprang at it, smashing his club into the serpent's heads. But every time he struck off one of the heads, two new heads grew back in its place.

Iolaus then set the surrounding forest on fire and with flaming brands, burned off the stump of each severed head, until only the immortal head remained. Hercules cut this off and buried it under a rock. Then he dipped his arrows in the hydra's blood, making them deadly.

On star maps, Hydra is depicted carrying two other constellations on its tail. They are Corvus (the Crow) and Crater (the Cup), and are important in another story. Apollo sent the crow to fetch water in a cup from a spring. On the way, it spied a fig tree and waited for days for the fruit to ripen, then gorged itself. Knowing that it would incur Apollo's wrath for such a delay, it snatched the water serpent in its claws and flew back to Apollo with it, explaining that the serpent had prevented it from filling the cup.

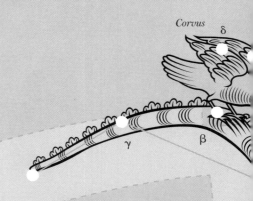

Corvus

δ

γ

β

One solitary star

Big though the constellation is, Hydra
has little to offer the casual observer.
The only star of note is a second
magnitude Alpha (α), named Alphard,
meaning the solitary one. This is an apt
description because it lies in a barren
region of the heavens. It is best found
by reference to the bright Regulus in
Leo and Procyon in Canis Minor, with
which it forms a triangle.

The quadrilateral grouping
of stars that forms Corvus, at
the tail end of Hydra, looks good
viewed through binoculars. These
stars can best be located with
reference to the nearby brilliant
Spica in Virgo.

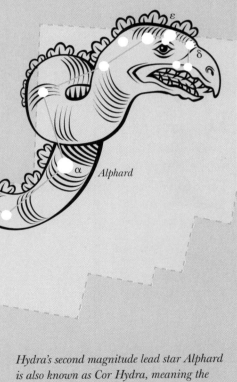

*Hydra's second magnitude lead star Alphard
is also known as Cor Hydra, meaning the
hydra's heart.*

Leo
The Lion

This constellation is one of the few that resembles the figure it is named for. With only a little imagination, one can see in the pattern of its stars the figure of a crouching lion. The lion constellation was revered in ancient Egypt, for the Sun entered it at the time of the flooding of the Nile River.

In Greek mythology, Leo was the Nemean lion Hercules fought with in the first of his labors. It was sent from the Moon by Hercules' stepmother and mortal enemy, Hera. The lion lived in a cave, from which it emerged to prey on the local people. When Hercules came upon the lion, he attacked it with spears and arrows, but they all bounced off—its skin was invincible. Hercules had to resort to hand-to-hand combat. He managed to strangle the beast. Then he skinned it and made its skin into a cloak to make him invincible, too.

The lion's heart

Leo is the one of the most recognizable constellations in the heavens. The curve of stars that make up the lion's head and forelimbs resembles an old-fashioned sickle used for reaping, and so is named the Sickle. At its southern end, is the brightest of Leo's stars, Regulus. It is also known as Cor Leonis, or the Lion's Heart.

Leo, and in particular the Sickle, is the apparent source of one of the most spectacular meteor showers. Called the Leonids, they may rain down on the Earth at the rate of hundreds an hour around November 17 each year.

What the astrologers say

Leo is a constellation of the zodiac, lying between Cancer and Virgo. Currently, the Sun passes through Leo between August 10 and September 16. In astrology, Leo is the fifth sign of the zodiac, covering the period from July 23 to August 22. According to astrologers, Leos can be generous and loyal, but they can also be egocentric and display a violent temper. Frequently theatrical, they are found widely among actors and film directors.

The Leonids—showers of meteors that rain down in November each year—were particularly heavy in 1966 and spectacular in 1999. This is because the source of the meteors, the comet Tempel-Tuttle, had just passed by.

Crouching Leo is one of the most recognizable constellations.

Libra

The Scales

Libra is one of the fainter constellations, that suffers by comparison with its dazzling neighbor Scorpius (see pages 88–89). But, being so located, it is an easy constellation to find. The association of Libra with scales or balance—and by extension harmony and justice—dates to ancient Babylonian times. Then, the autumnal equinox occurred in Libra, when the days and night are of equal length.

The Greeks did not consider Libra to be a separate constellation, but saw its stars as part of Scorpius, in particular the Scorpion's claws. It was the Romans who began to identify Libra with scales again, appreciating the balance of the days and nights at the equinox. Libra is adjacent to Virgo (see pages 96–97), sometimes pictured holding the scales and identified as Astraea, goddess of justice.

The scorpion's claws

Astronomically, there is nothing spectacular in Libra. Its two brightest stars Alpha (α) and Beta (β) are just third magnitude and make a noticeable triangle with the fainter Gamma (γ). Alpha and Beta have delightful Arabic names—Zubenelgenubi (meaning the southern claw) and Zubenelchemale (northern claw).

What the astrologers say

Libra is a constellation of the zodiac, lying between Virgo and Scorpius. Currently, the Sun passes through Libra between October 31 and November 23. In classical times, it used to be the constellation of the autumnal equinox, but this is now in Virgo.

In astrology, Libra is the seventh sign of the zodiac, covering the period from September 23 to October 22. Librans are reputed to be well-balanced, as their sign might indicate. They are quiet extroverts, but can be indecisive. Affectionate and romantic, they are rarely short of sexual partners.

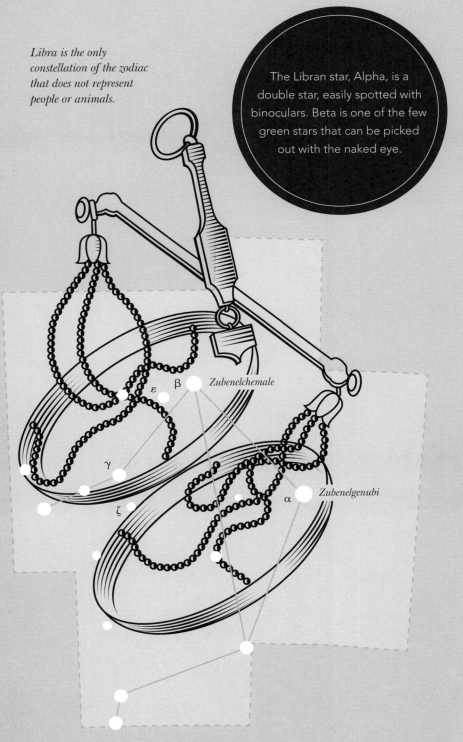

Libra is the only constellation of the zodiac that does not represent people or animals.

The Libran star, Alpha, is a double star, easily spotted with binoculars. Beta is one of the few green stars that can be picked out with the naked eye.

ε   β   *Zubenelchemale*

γ

ζ   α   *Zubenelgenubi*

Lyra
The Lyre

This tiny constellation, sandwiched in between Cygnus and Hercules, was sometimes pictured as a bird of prey. Indeed, the name of its brilliant star Vega, means swooping eagle in Arabic. But in mythology Lyra was the lyre. The messenger of the gods, Hermes (Mercury in Roman mythology) made the first lyre, it is said, out of the shell of a tortoise and strung it with strings made of cow gut.

Hermes gave the lyre to Apollo, whose son Orpheus learned to play it with such beauty that the wild beasts came to listen; even the trees followed him. Orpheus sailed with the Argonauts on their quest for the Golden Fleece. With his beautiful singing, he lulled the guardian dragon to sleep and later outsung the Sirens, whose seductive voices would have tempted the Argonauts into the sea to be drowned.

But the most famous tale of Orpheus relates to his love of the fair nymph Eurydice, whom he married. One day, while she was escaping the advances of another of Apollo's sons, Aristaeus, she was bitten by a poisonous snake and died. Orpheus felt he couldn't live without Eurydice, and descended into the Underworld to bring her back to Earth.

He pleaded with Hades, the god of the Underworld, to let her go, and Hades eventually agreed, persuaded by Orpheus' beautiful music. But there was one condition: Orpheus must never look at his beloved Eurydice until they had returned to the surface. They had almost reached the gates of Hades when Orpheus looked round to see if Eurydice was still with him. Immediately, she vanished back into the Underworld forever.

The Harp Star

The lead star in Lyra, Vega, is often called the Harp Star. Of first magnitude, it is the fifth brightest star in the heavens. It shines brilliantly high overhead in summer in the Northern Hemisphere, when it forms one corner of the conspicuous Summer Triangle of stars. The others are Deneb in Cygnus and Altair in Aquila. Lyra's other gem, between the stars Beta (β) and Gamma (γ), is a patch of gas known as the Ring Nebula (M57), which looks exquisite through a large telescope.

Thanks to the slight wobbling of the Earth's axis, known as precession, Vega in Lyra will become the Pole Star in about 22,000 years' time.

Arab astronomers identified Lyra with a bird of prey. The name Vega is derived from the Arabic meaning swooping eagle.

Orion
The Mighty Hunter

Orion sits on the celestial equator, which makes it equally visible to observers in both the Northern and Southern Hemispheres. It requires only a little imagination to convert the pattern of bright stars into the figure of a mighty hunter. He has his right arm raised, ready to strike a blow with a brutal-looking club, while his left arm carries a shield.

A sword hangs from his belt, marked by a diagonal composed of three bright stars. In mythology, Orion was the son of Poseidon, the god of the sea, and Euryale, the daughter of King Minos of Crete. Poseidon gave Orion the power to walk on water. Orion went hunting with two dogs at his heels, represented in the heavens by Canis Major (the Great Dog) and Canis Minor (the Little Dog).

Orion had enormous stature and possessed prodigious strength. He was also handsome. Unsurprisingly, many of the stories about him concern his love of beautiful women. In one, he was chasing the seven beautiful sisters known as the Pleiades. Just as he was about to catch them, Zeus intervened, turned them into doves, and later placed them among the stars. And this scenario is re-enacted in the heavens, where Orion stills pursues them.

Orion's riches

The constellation's two first magnitude stars, Betelgeuse and Rigel, both supergiants, provide a nice contrast. Rigel is brilliant white while Betelgeuse is noticeably red. Another delight in Orion is the bright patch visible to the naked eye beneath the three stars marking Orion's Belt. Small telescopes show that this patch is actually a bright nebula, a glowing mass of gas and dust. Telescopes are needed to show its true magnificence and kaleidoscope of colors. It is aptly named the Great Nebula in Orion. Astronomers know it as M42.

The Orion Nebula is a vast star-forming region that has been explored extensively by the Hubble Space Telescope. Nearly the whole of Orion is, in fact, embedded in gas and dust. And near the star at the lower end of Orion's Belt is another famous nebula, the Horsehead—a dark nebula bearing an uncanny resemblance to a horse's head.

Orion Nebula can look spectacular through a telescope.

Orion's star Betelgeuse is supergigantic, being one of the biggest stars known, with a diameter of some 250 million miles (400 million km), over 250 times bigger than our own Sun.

α

Betelgeuse

γ

δ

ε

ζ

M42

β

Rigel

Pegasus
The Winged Horse

The winged horse was a favorite concept in the ancient world. It was featured prominently in the art of the ancient Assyrian civilization around 1500 B.C.E.

It appeared on Egyptian coins of the same era and later on Greek coins. In Greek mythology, the flying horse, known as Pegasus, was placed among the constellations by Zeus himself.

It was Zeus' son Perseus who was responsible for Pegasus' birth. When this famous hero beheaded the snake-haired Medusa (see pages 82–83), Pegasus was born from the blood spurting out of her dying body. This fabulous creature spread its wings and soared into the sky to join the nine Muses on Mount Helicon. They were Zeus' daughters, and goddesses of the arts and sciences.

The goddess Athene came to see Pegasus and tamed it with a golden bridle. She later gave the bridle to the hero Bellerophon so he could ride Pegasus when he went to fight the Chimera. This was a ferocious, fire-breathing monster, with the head of a lion, the shaggy body of a goat, and the tail of a dragon.

Having killed the Chimera, Bellerophon went on to conquer other adversaries and, proud of his achievements, flew into the sky on Pegasus to join the gods on Mount Olympus. Pegasus threw him off, and he fell back to Earth, but Pegasus went on to join the gods and serve Zeus himself by carrying his thunderbolts.

The great square

The constellation Pegasus is a dominant feature of the autumn skies in the Northern Hemisphere. It is easily recognized by its almost perfect square of four stars, which form the famed Great Square of Pegasus. The square stands out clearly in what is otherwise a relatively starless part of the heavens. All four stars in the square are equally bright, of about the second magnitude. Only three of the stars belong to Pegasus, however: Beta (α), Alpha (β), and Epsilon (ε). The other is the lead star of the linked constellation Andromeda. With small telescopes, Epsilon, also known by its Arabic name, Enif, proves to be a double star with a faint companion.

It is interesting to compare the colors of the Great Square stars, Beta, Alpha, and Epsilon. They are noticeably different —red, white, and yellow respectively.

The Great Square acts as a good signpost to fainter constellations, such as Aquarius and Pisces.

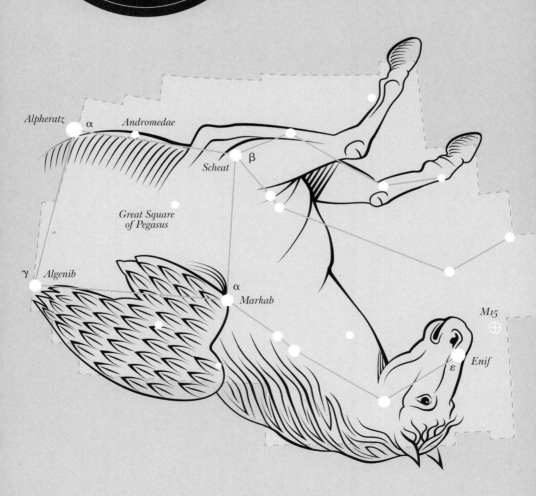

Alpheratz α   Andromedae

β

Scheat

Great Square
of Pegasus

γ Algenib

α
Markab

M15

ε Enif

Perseus
Slayer of Monsters

Perseus was one of the legendary Greek heroes, like Hercules (see pages 68–69) and is featured in many an adventure. He is immortalized in the heavens next to his beloved Andromeda.

Perseus led an eventful life right from conception. His mother Danae, had been imprisoned by her father in an underground cell, lit only by a barred window in the ceiling. But the king of the gods, Zeus, desired her, and went to her in a shower of gold that fell as rain through the window. She became pregnant and gave birth to a boy-child, Perseus. Mystified and furious, her father locked them in a chest and cast them into the sea. To make a long story short, a fisherman saved them.

Later Perseus was tricked into going after the head of Medusa, one of the hideous Gorgon sisters. Medusa had snakes for hair, and her gaze would turn people into stone. But the gods were with Perseus, giving him a shining shield, a helmet that made him invisible, a sword, and winged sandals so that he could fly. Thus equipped, he flew to where the Gorgons lived. Finding Medusa, he looked at her only as a reflection in his shield. Then he sliced off her head and flew away with it. It was on his way home that he came across Andromeda chained to the rocks, and killed the sea monster, Cetus (see pages 54–55), that was about to eat her. He later married Andromeda, who bore him six children.

The winking demon

In the sky, Perseus is depicted holding Medusa's head. The constellation lies within the Milky Way and so is a delight to sweep with binoculars. The brightest star, Alpha, is also known by its Arabic name Mirphak.

Perseus's second brightest star, Beta, is located in Medusa's head. Its Arabic name is Algol, meaning demon's head, and it has long been known as the Demon Star. For most of the time, Algol shines steadily at a magnitude of about two. But about every three days, it dims noticeably. It is a kind of variable star known as an eclipsing binary. A binary is an object that looks like one star, but is actually two, orbiting around each other. In Algol's case, one star is large and dim, the other small and bright. About every three days as we view the pair from Earth, the large, dim one covers up the small bright one, and so we see the overall brightness fall. When the dim star passes by, the former brightness returns.

Between Mirphak and the distinctive W shape of Cassiopeia, lies a superb pair of star clusters known as the Double Cluster (also called the Sword Handle). They are just visible to the naked eye and look magnificent when viewed through binoculars.

With a mighty blow, Perseus beheaded the snake-haired monster Medusa.

Pisces
The Fishes

Perhaps surprisingly, this faint, sprawling constellation is one of the most ancient, and has always been associated with fish. As depicted in the heavens, the two fishes are swimming in opposite directions and have their tails joined by a cord. In mythology, the two fishes represent Aphrodite (Venus) and her son, Eros (Cupid). One day they had to hide in the rushes along the bank of the Euphrates River to escape the awesome dragon-headed monster Typhon. When the monster was nearly upon them, two fishes swam up and carried them away to safety, becoming immortalized in the sky as the constellation Pisces.

Watery and faint

Pisces lies in an undistinguished region of the heavens, populated by other faint "watery" constellations, such as Cetus and Aquarius. Its brightest stars are only just above the fourth magnitude.

The only sure way of locating Pisces is by reference to Pegasus's famous square. One of the two fishes lies due south of the square, outlined by a circle of stars sometimes called the Circlet of Pisces. In this circlet is the star TX, which binoculars show to be a striking brick-red color.

What the astrologers say

Pisces is a constellation of the zodiac, lying between Aquarius and Aries. Currently, the Sun passes through Pisces between March 12 and April 18. The point of intersection between the path of the Sun through the heavens (the ecliptic) and the celestial equator takes place on about March 20. It is known as the First Point in Aries, because in classical times, the Sun crossed the Equator in Aries. However, because of precession—the slight gyration of the Earth's axis, the Sun now crosses in Pisces.

Astrologers ignore this fact, and in astrology Pisces, the twelfth sign of the zodiac, still covers the period from February 19 to March 20. The Piscean character, astrologers say, is complex and enigmatic. Pisceans are sensitive, and often artistic and creative. They tend to fall in and out of love readily.

α

*Located near Eta, M74 is a
face-on spiral galaxy with
wide-open arms.*

The vernal (spring) equinox
falls on about March 20 when the
Sun crosses the Equator traveling
north. This signals the beginning
of spring in the Northern
Hemisphere.

*Square of
Pegasus*

M74   η

δ

Circlet

TX

γ

β

Sagittarius
The Archer

In the heavens, we see Sagittarius as a centaur, half-man, half-horse. He is the consummate archer, keen of eye and with a deadly aim. Sagittarius dates back to Babylonian times, when the centaur was a favorite creature. Greek historians suggest that Sagittarius was a two-legged, satyr-like beast, half-man, half-goat. Sagittarius was supposed to be Crotus, inventor of archery, who was fathered by the pipe-playing god Pan.

In the center of things

Sagittarius lies in the direction of the center of our galaxy, so when you look at it, you are peering into the densest possible concentration of stars. Possibly because of its location, Sagittarius is not the easiest constellation to identify. It is best to first find the adjacent constellation Scorpius. The six bright stars that outline the Archer's bow and arrow then become evident. The arrow, with Gamma (γ) at the head, points menacingly at Antares, the orange-red star that marks the Scorpion's heart.

The Lagoon nebula, also known as M8, is just visible to the naked eye near the upper part of the bow. But it is best seen in binoculars or a small telescope. It makes a rough triangle with Mu (μ) and Lambda (λ). The Trifid (M20) lies slightly farther north and appears as a misty patch in binoculars. But you'll need a telescope to see its three dark dust lanes.

What the astrologers say

Sagittarius is a constellation of the zodiac, lying between Scorpius and Capricornus. Currently, the Sun passes through Sagittarius between December 18 and January 19. In astrology, Sagittarius is the ninth sign of the zodiac, covering the period from November 22 to December 21. Sagittarians are supposed to be chatty, outgoing, and lively people, even hyperactive. Ever adventurous, they have a fiery temperament and fall in and out of love impetuously.

Arab astronomers named
Alpha, Rukbat, meaning
"knee of the archer."

Sagittarius is one
of the most spectacular
constellations. It lies in the richest
region of the Milky Way and is
awash with nebulae and clusters.
Two of the most famous nebulae
of all lie in Sagittarius:
Lagoon and Trifid.

π

σ

μ

M20

M8

λ

δ

Galactic
Center

γ

ε

α

Rukbat

β

Scorpius
The Scorpion

Scorpius is one of the few constellation patterns that truly resembles the figure it is meant to represent. Only a little imagination is required to link up the bright stars into the figure of a deadly scorpion, with its curved tail poised ready to sting.

Scorpius is one of the oldest constellations, recognized by the Babylonians in the Euphrates valley 7,000 years ago, and associated with their god of war. In ancient Greece, it was a much larger constellation than we see, as it included the Scorpion's claws, which form part of the constellation Libra. In Greek mythology, the Scorpion was the creature that killed the famed hunter, Orion.

Orion was the mightiest of hunters. Unwisely, he boasted to Artemis, the goddess of the hunt, that he could track down and kill any creature on the face of the Earth. With this, the Earth trembled with rage, and cracked open. A scorpion scuttled out and stung Orion to death. With some compassion, the gods placed Orion and the Scorpion on opposite sides of the heavens, so that Orion sets in the west as the Scorpion rises in the east.

Rival of Mars

The brightest star in Scorpius is Antares. It is a huge supergiant star that is noticeably orange-red. Its name means "rival of Mars," for its color is similar to that of the Red Planet, Mars. Close by Antares is a magnificent globular cluster, M4, which is just visible to the naked eye and easily seen in binoculars. But a telescope is needed to bring into focus its individual stars.

What the astrologers say

Scorpius is a constellation of the zodiac, lying between Libra and Sagittarius. Currently, the Sun passes through Scorpius only briefly, between November 23 and 29.

Astrologers always call this eighth sign of the zodiac Scorpio. They say it covers the period from October 23 to November 21. Scorpios, astrologers tell us, tend to be intense in everything they do, and intensely emotional. They can be easy-going, but can become testy if they are pushed too far (remember the venomous sting in the tail!). With typically warm, magnetic personalities, they make marvelous lovers.

Down at the tail end of the Scorpion, the region around Zeta (ζ) is packed with interest. Zeta itself is a colorful double star, and just north of it is a brilliant open star cluster NGC6231, full of young hot stars. Lambda (λ), or Shaula, marks the Scorpion's sting.

Scorpius is one of the dazzling far southern constellations that northern astronomers can only glimpse.

β

δ

Antares

α   M4

ε

λ

Shaula

6231

ζ

Taurus
The Bull

In the heavens, Taurus depicts the front part of a bull, with a wicked red eye and horns lowered ready to charge. In mythology, the Bull represents Zeus in disguise pursuing a fair maiden, the beautiful Europa. Europa was the daughter of Agenor, king of Phoenicia.

Europa was playing one day on the seashore with her girl friends when she spied a beautiful white bull grazing quietly among her father's herd. Not knowing, of course, that it was really Zeus, she stroked it and climbed on its back. The beautiful bull then leapt into the sea and swam, with the now terrified Europa on his back, to Crete. There, Zeus revealed his true self and made love to her. Among the three children born of the union, one was Minos, who became king of Crete and who established bull worship at his palace at Knossos. There, too, he kept his monstrous offspring the Minotaur, half-man, half-bull, that inhabited the Labyrinth and lived on human flesh.

The red eye

Taurus is a splendid constellation, close to Orion in the heavens. The three stars that make up Orion's Belt act as pointers to the supergiant star Aldebaran, the baleful red-orange eye of the Bull. Aldebaran is located among a V-shaped cluster of fainter stars known as the Hyades, although it actually lies much closer to us.

What the astrologers say

Taurus is a constellation of the zodiac, lying between Aries and Gemini. Currently, the Sun passes through Taurus between May 14 to June 21.

In astrology, Taurus is the second sign of the zodiac, covering the period from April 20 to May 20. Taureans are reputed to be laid-back characters, steady, reliable, and practical. They value wealth and status. Though not particularly romantic and slow to form relationships, they make endearing and tolerant partners.

Continuing the line through Orion's Belt and Aldebaran brings you to the Pleiades (M45), the best open cluster in the whole heavens. It is known popularly as the Seven Sisters, but not even the keenest-sighted people can make out its seven brightest stars.

The V-shaped group of stars near Aldebaran form an open cluster called the Hyades.

Ursa Major
The Great Bear

Ursa Major is a sprawling constellation that occupies a huge area of the northern skies: it rates third in size among the constellations. It is best known, not for its bear shape, but for the pattern made by its seven brightest stars. The pattern resembles both the handle and share (the part of the plow that cuts the earth) of an old-fashioned plow, and the shape of a ladle for dipping into buckets of milk or water.

The star group is called the Big Dipper—"Big" because there is a smaller dipper (the Little Dipper) close by. For much of North America and Europe, the Big Dipper is always visible in the night sky because it lies not far from the north celestial pole—it is one of the circumpolar constellations.

To the ancient Greeks, the Great Bear represented one of Zeus' many conquests, the nymph Callisto. The daughter of Lycaon, king of Arcadia, Callisto became a companion of the virgin god Artemis and agreed to lead a life of chastity. Unfortunately for her, Zeus came by one day and became captivated by her extraordinary beauty.

He transformed himself into Artemis to get close to her as she was resting. As Callisto and the false Artemis embraced, Zeus turned himself back into his usual form and promptly seduced her.

Callisto became pregnant and was banished by Artemis. She gave birth to a son, Arcas. From here on, there are several different versions of how Callisto came to be turned into a bear. One says Zeus' wife Hera did it, enraged by her husband's philandering. Later, Arcas came across the bear-like Callisto when out hunting and would have killed her. But Zeus came to her rescue and sent a whirlwind to whisk them into the heavens as the constellations Ursa Major and Boötes.

The Pointers

All of the stars in the Big Dipper have names. Merak and Dubhe in the share are known as the Pointers because a line drawn through them points in the direction of the Pole Star, Polaris. This fact has helped navigators through the centuries because it is a sure way of finding north.

Among the Big Dipper stars, Mizar is the most interesting. To the naked eye, it is a double star, with a fainter companion, Alchor. Through a telescope, Mizar itself appears double, too. And by checking the spectra of each of its components, these also turn out to be doubles. So we can think of Mizar as a double, double double.

All the Big Dipper stars, except Alkaid and Dubhe, travel through space together, forming a moving cluster. α and β act as pointers to Polaris (Pole star).

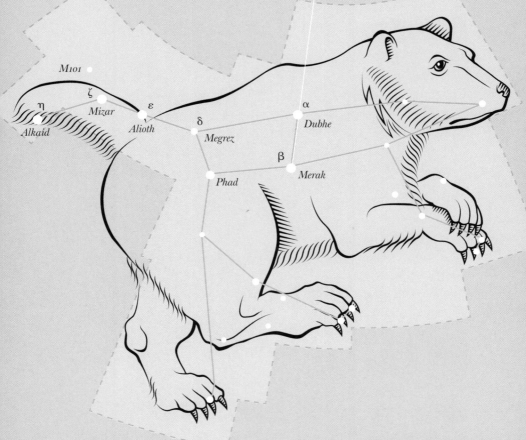

M101

ζ

η   Mizar   ε

Alkaid   Alioth   δ   α

Megrez   Dubhe

β

Phad   Merak

Ursa Minor
The Little Bear

The Little Bear is much less conspicuous than the Great Bear. Its main star pattern imitates the shape of the Big Dipper, and is often called the Little Dipper. The tip of the Little Bear's tail is marked by the constellation's brightest star, Polaris. This is also known as the Pole Star and North Star because it lies close to the north celestial pole of the heavens, around which all the stars appear to revolve during the night.

The Greek philosopher Thales of Miletus reportedly introduced the Little Bear as a constellation in the sixth century B.C.E. as an aid to navigation.

In mythology, the Little Bear and the Great Bear have been associated with the story of the birth of Zeus on Crete. (Other myths surround the Great Bear, see pages 92–93.)

Zeus' mother Rhea had fled to a cave in Crete to escape from his father, Cronus, who had eaten all their other children because of his fear that they might one day overthrow him. On Crete, Zeus was nursed by the nymphs Adrasteia and Ida. They made the baby a beautiful golden ball to play with that whooshed through the air like a shooting star. Cretan warriors guarded the entrance to the cave, clashing their swords and shields to drown the baby's cries so that Cronus could not hear him. When Zeus grew up, and did indeed overthrow his father, he placed Adrasteia in the sky as the Great Bear and Ida as the Little Bear.

At the tip of the tail

Polaris, at the tip of the Little Bear's tail, is the Pole Star—for the present. It lies within a degree of the celestial north pole, and is edging ever closer—it will be closest in the year 2095. But Polaris has not always been the Pole Star, nor will it be in the future. Because of a slight gyration of the Earth's spin axis, the north celestial pole describes a circle in the heavens, coming full circle every 26,000 years or so. In about 10,000 B.C.E., the star Vega in Lyra was pole star. Around 3,000 B.C.E., it was Thuban, the lead star in Draco.

Polaris, the Pole Star, is a second magnitude star, best located using two of the stars in the Big Dipper as pointers. Polaris is also a double star, one component of which is a Cepheid—a variable star that varies in brightness like clockwork.

The location of Polaris near the north celestial pole has proved a boon to navigators for centuries.

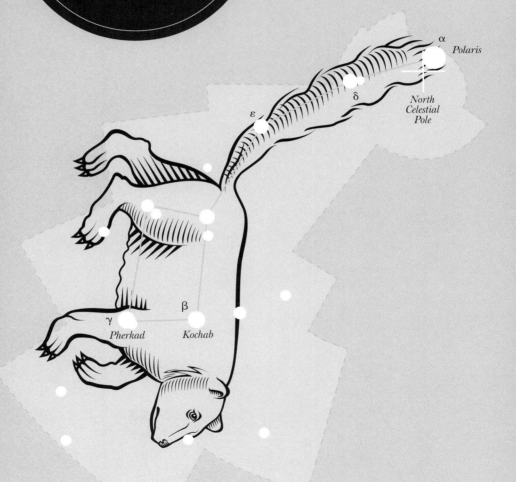

α
Polaris

δ

North
Celestial
Pole

ε

γ
Pherkad

β
Kochab

Virgo
The Virgin

The Virgin is depicted in the heavens with wings, holding an ear of wheat. In the ancient world, she was associated with the great goddess of the harvest, since the Sun passed through the constellation around harvest time. In ancient Babylon, she was Ishtar, a woman of immense sexual appetite who led a cult of sacred prostitution.

In ancient Greece, Virgo was Demeter, the goddess of agriculture and fertility. By Zeus, she gave birth to a daughter, Persephone. Hades, god of the Underworld and Zeus's brother, kidnapped Persephone and took her to his underground world and made her his wife. Demeter searched far and wide for Persephone, neglecting the crops she was supposed to nurture. Eventually she learned what had happened, and Zeus persuaded Hades to return Persephone to the living Earth. But Persephone has to go back to the Underworld for part of every year because she ate some pomegranates while she was there. When she goes back each year to the Underworld, winter descends on Earth and crops die. When she returns in the spring, the Earth becomes alive and fruitful again.

The great cluster

Virgo is disappointing to the naked eye. Only its lead star, Spica, is bright—of the first magnitude. The main interest in Virgo lies in its deep-sky objects, in the multitude of distant galaxies visible in telescopes. These galaxies form part of an enormous group of galaxies called the Virgo cluster.

What the astrologers say

Virgo is a constellation of the zodiac, lying between Leo and Libra. Currently, the Sun passes through Virgo between September 10 and October 31. In astrology, Virgo is the sixth sign of the zodiac, covering the period from August 23 to September 22. Virgos are supposed to be intelligent and modest, but they can become fussy and critical. They tend to be serious people given to worry. Overtly they are cold and unaffectionate, but this often conceals a passion within.

Virgo is a sprawling constellation that is second in size only to Hydra. There are probably as many as 3,000 galaxies in the Virgo cluster, lying up to about 70 million light-years away.

Bright star Spica marks the virgin's left hand.

Virgo cluster

ε

δ

β

ζ

Porrima

γ

ι

α

Spica

The Wandering Stars

Among the brightest objects in the night sky are the stars that wander against the background of the fixed stars in the constellations. Look at these wandering stars through binoculars or a small telescope, and you will find that they are not like the other stars. No matter how powerful your binoculars or telescopes are, the fixed stars appear only as tiny pinpricks of light. But the wandering stars present a distinct disk. They are not remote bodies like the fixed stars, but relatively close neighbors of the Earth in space.

The Wandering Stars
Sun, Moon, the planets, and more

Two thousand years ago, the Greeks had a word for the five wandering stars they had discovered—they called them planets. Today, we know them not by their Greek names, but by their Latin names—Mercury, Venus, Mars, Jupiter, and Saturn.

The Greeks thought that the planets circled around Earth. But in fact they orbit around the Sun, just like Earth does—Earth is a planet, too. There are two other planets, but they are so far away that we can only see them only through a telescope. They are Uranus and Neptune. This makes eight planets in all. The planets form the major part of the Sun's family in space, or the solar system. Other members of this family include the satellites, or moons, that circle around many of the planets. Then there are a number of "dwarf" planets, such as Pluto. And

there is a swarm of "miniplanets" or asteroids, found between the orbits of Mars and Jupiter.

Comets also belong to the Sun's family. These tiny icy bodies journey in toward the Sun from the remote depths of the solar system and begin to shine only when they near the Sun and start to melt. At their brightest, they can become the most spectacular objects in the heavens.

The scale of the solar system is awesome. Earth is one of the inner planets that lie quite close together in the heart of the solar system. Yet it is still 93 million miles (150 million kilometers) from the Sun. The outer planets lie much farther apart, separated by thousands of millions of miles. The distant dwarf planet Pluto wanders some 5 billion miles (7 billion kilometers) away from the Sun.

A depiction of the appearance of Halley's Comet in 1546.

The most brilliant of all the wandering stars is the sparkling evening star, which hangs in the western sky just after sunset on many nights of the year.

The Sun

To us, the Sun is the most important of all the heavenly bodies, pouring light and heat onto our world. Sunlight is needed to make plants grow, and without plants for food, we and other animals could not survive. The Sun's heat warms our world and provides comfortable conditions for us and millions of other species of living things to thrive and multiply. Small wonder, then, that ancient peoples worshiped the Sun as a powerful god.

The Babylonians and the Assyrians in Mesopotamia were Sun worshipers, as were the ancient Egyptians. To the Egyptians, the Sun god Re (or Ra) became all important. They thought that he sailed across the sky every day carrying the Sun in a boat. They pictured him with a human body and a falcon's head. The pharaohs came to consider themselves the sons of Re. Heliopolis ("Sun City") in Lower Egypt became the center of Sun worship.

In ancient Greece, Helios was the Sun god. It was he who drove the Sun chariot across the sky every day. He was especially worshiped in Rhodes. From about 500 B.C.E., however, Apollo, the god of light and purity, became increasingly identified as a Sun god. Apollo was the son of Zeus and Leto, and the twin of Artemis, who became goddess of the hunt. He was a versatile and powerful god, a celebrated musician and poet, archer, and healer.

In later times, Sun worship dominated the religions of three great civilizations in Central and South America—the Mayas, Incas, and Aztecs. Human sacrifice featured heavily in their rituals. The Aztecs, for example, believed that their Sun god Huitzilopochtli would die unless he was offered human blood and hearts every day.

The Sun is a very ordinary star. It just seems bigger and brighter than other stars because it is very much closer to us. As stars go, it is not particularly big or bright—there are supergiant stars that are hundreds of times bigger and tens of thousands of times brighter. In fact, astronomers class the Sun as a dwarf—a yellow dwarf because the light it gives off is a yellowish color.

Apollo was worshipped as the Sun god in ancient Greece and Rome.

The Moon circles around Earth in just 27.3 days, and it also spins on its own axis in the same period. As a result of these motions, the Moon always presents the same face toward us—the nearside.

The Greek goddess Artemis was the sister of the Sun god, Apollo.

The Moon

The Moon is Earth's constant companion and closest neighbor in space. It is the only other world that humans have set foot on and explored—so far. Apollo astronauts made six landings on the Moon between 1969 and 1972.

The Moon is Earth's only natural satellite, circling around Earth once a month. During this time we see it appear to change in shape—go through its phases—from slim crescent to full circle and back again. It lightens our world by night, just as the Sun lightens our world by day.

Like the Sun, the Moon was widely worshiped in the ancient world. In early Greek mythology, the Moon goddess was Selene, also called Mene. She was the sister of the Sun god Helios. Every evening, as Helios finished his journey across the daytime sky, Selene started hers in the nighttime sky. Later, the goddess Artemis (Diana in Roman mythology) took on the role of a Moon goddess. She was a divinity of the light like her brother Apollo, who became a Sun god. Artemis was also goddess of the hunt and wild animals.

Artemis was the ultimate virgin goddess who gathered around her a band of maidens vowed to chastity. Those who lost their virginity even by trickery (like Callisto) were banished in disgrace. And woe betide any mortal Peeping Tom, such as the lovelorn Actaeon, who happened upon Artemis bathing naked. Artemis changed Actaeon into a stag and then set her hounds upon him, that tore him limb from limb and then gobbled him up.

A dead world

The Moon is a rocky body like Earth, but is a very different world. It is much smaller than Earth—only about one-third of the size across. Being so small, it has a small mass and low gravity. As a result, the Moon could not retain any atmosphere. With no atmosphere, there are no clouds on the Moon, no rain, and no blue sky.

With our eyes we can make out two different kinds of regions on the nearside, dark and light. Through binoculars and telescopes we can see that the darker regions are great plains, which early astronomers called seas. And the lighter regions are highlands. We can see craters everywhere, some measuring hundreds of miles across.

Mercury

Mercury is the planet closest to the Sun, and the one that travels the fastest. Not an easy planet to spot, it can either be seen low down near the eastern horizon before dawn or low down near the western horizon after sunset. At its brightest, it is slightly more brilliant than Sirius, the brightest star.

Ancient astronomers were familiar with Mercury when it appeared in the morning and the evening, but they did not realize it was the same body. When they saw it as a morning star, they called it Mercury, but called it Apollo as an evening star.

Mercury is an appropriate name for this fast-moving planet, for Mercury was the fleet-footed messenger of the gods in Roman mythology. He is usually depicted as a lithe, handsome youth with a winged helmet and winged sandals so that he could fly swiftly to do the gods' bidding. In one hand he carries a magic wand called the caduceus, which has wings at the top and snakes twined round it.

Mercury was known as Hermes in Greek mythology and was the son of Zeus and Maia, a daughter of Atlas—the Titan condemned forever to hold the heavens on his shoulders. On the very day he was born, Hermes displayed his talents for mischief and invention. He stole cattle from Apollo and constructed a new musical instrument, the lyre. He subsequently presented it to Apollo to make amends for the theft of his cattle.

Another moon

Mercury is the smallest planet. It is less than half as big across as Earth. As a result it has a low mass and low gravity and has been unable to hold onto any atmosphere. Being close to the Sun, we would expect it to be hot, and it is. Parts of the planet facing the Sun reach temperatures up to 800°F (430°C)—which is hot enough to melt lead. Temperatures on the opposite side of Mercury facing away from the Sun plummet to -290°F (-180°C).

Mercury (Hermes), the fleet-footed messenger of the gods.

Mercury is covered with craters, and looks much like some parts of the Moon. These craters were dug out by meteorites billions of years ago. A particularly big one created a huge basin called the Caloris Basin. Some 800 miles (1,300 kilometers) across, it is the most prominent feature on the planet.

Venus

Of all the planets, Venus is the easiest to spot. It is by far the brightest—far brighter than any of the stars. It hangs in the early evening sky on many nights of the year, in the west just after sunset. It is the evening star. At other times of the year, early risers may also see Venus as the morning star, hanging in the east as the sky brightens just before sunrise.

The Babylonians associated the planet with Ishtar, their goddess of beauty, fertility, and war. The Romans named it after their goddess of love and beauty, Venus. Of obscure origins, Venus was originally connected with fruitfulness and crops. It was only later that she took on the attributes of the Greek goddess Aphrodite. One story tells that she was a daughter of Zeus, another that she was born from the foam of the ocean. Aphrodite was the essence of beauty, whose figure and features were perfection.

As might be imagined, the other goddesses on Olympus were not well pleased when the beautiful Aphrodite took her place among them. Her main rivals were Hera and Athene. One day all three were arguing about which one was the fairest, when Zeus intervened and said that a mortal must make the decision.

Paris, the son of King Priam of Troy, was made to choose. To help him make up his mind, Hera promised that he would be lord of all Asia if he should judge her the most beautiful; Athene, that he would be invincible in battle. Aphrodite promised that she would help him win the most beautiful mortal woman. There was no contest—he chose Aphrodite. And with Aphrodite's help, he carried off Helen, wife of Menelaus, the king of Sparta, a powerful city-state in Greece. His refusal to hand her back sparked the Trojan War, in which the Greeks gained victory by using the Trojan horse.

Aphrodite's beauty stirred all of the gods, but strangely, she took the ugliest among them, Hephaestus (Vulcan in Roman mythology) as a husband. But it was inevitable that she should console herself with some of the more fetching of the gods, including Ares (Mars) and Hermes (Mercury). She also loved two mortals, Adonis and Anchises. From the union with the latter, she gave birth to Aeneas, one of the few to survive the fall of Troy. His wanderings thereafter were the subject of one of the world's great literary epics, the *Aeneid*, written by the Roman poet Virgil.

Venus, one story tells, was born from the foam of the oceans.

Venus is a near twin of Earth in size, is made up of rock, and has an atmosphere. The temperature on Venus reaches an amazing 900°F (480°C). The atmosphere is responsible for this, acting like a greenhouse to trap the Sun's heat. Most of Venus's surface consists of low-lying plains, with a few highland areas dotted about, and has been shaped almost entirely by volcanoes.

In Greek mythology Gaea is the Mother Earth goddess, ancestral mother of all life.

Subtle forces of change are at work on Earth: the forces that cause erosion. The action of the weather is a powerful erosive force. The Sun's heat, frost, and the sandblasting effect of dust-laden wind take their toll. Flowing water cuts into and dissolve rocks, depositing the debris elsewhere.

Earth

The planet we live on is the largest of the rocky planets. It is unique among these and indeed all the planets because it is a cradle of life. It has just the right conditions to allow an abundance of life to survive and thrive. Such conditions do not exist on any other planet—at least in our part of the universe.

Earth was personified as a deity in both Egyptian and Greek mythologies. To the Egyptians he was the god Geb, whose sister and wife was the sky goddess Nut. It was her star-studded body that was held over Geb by Shu, the Egyptian equivalent to Atlas.

The Greeks conceived Earth as a deep-breasted goddess, Gaea. She emerged out of the primeval emptiness they called Chaos. She bore Uranus, the god of the heavens. The Earth and the heavens—the universe—was created, but not peopled. So Gaea lay with Uranus to beget the first race of the gods, the Titans, and the one-eyed Cyclopes. She is also credited as giving birth to the human race of mortal men and women. Gaea had the gift of prophecy and was venerated at the Oracle at Delphi.

Continents adrift

Earth has the typical layered structure of all the rocky planets. It has a thin, hard crust, 70 percent of which is hidden by the water of the oceans. The crust overlays a deep rocky layer known as the mantle. At the Earth's center is a large core of iron and nickel, the outer part of which is liquid. Currents eddying inside the liquid core create electric currents and magnetism, turning Earth into a giant magnet, with a magnetic field extending into space.

The outer crust of Earth is not solid like an eggshell, but is split into sections known as plates. These plates are in constant motion over the face of the Earth, causing the oceans to widen and the continents to drift slowly apart. Movements along the plate boundaries set off earthquakes and trigger volcanic eruptions.

Mars

Mars cuts a distinctive figure in the night sky because it boasts a fiery reddish-orange color, which has earned it the title of the Red Planet. Like Venus, Mars is a neighbor of Earth in space, approaching at times to within 35 million miles (56 million kilometers). But it is much smaller than Venus, and therefore does not shine as brightly. Nevertheless, it still outshines all the stars in the sky when it gets closest to us and most brilliant.

Mars is named after the Roman god of war, probably because its reddish hue was reminiscent of fire and bloodshed. But Mars was originally a gentler god, of agriculture and fruitfulness. When Mars became a god of war, he took on the attributes and legends of the Greek god of war, Ares.

Ares was the son of Zeus and Hera, but was not popular among the other gods because of his brutality on the battlefield. He took the goddess of love and beauty, Aphrodite, as his wife, and she bore him three children—two sons (Phobos and Deimos) and a daughter (Harmonia). Phobos (Fear), and Deimos (Terror) accompanied Ares into battle, along with his sister Eris (Strife). Mars's two moons are named after Ares's two sons.

The myth of life

A century ago, many people believed that Mars was inhabited by an intelligent race of Martians. They pointed out similarities between Mars and Earth—it has ice caps; its day is only slightly longer than our own; "waves of darkening" sweep across the planet at times, which could signal the growth of vegetation. The Martian concept was triggered by the Italian astronomer Giovanni Schiaparelli in 1877, who reported seeing "canals" on the planet. People interpreted this as meaning artificial waterways, and thought intelligent beings lived there.

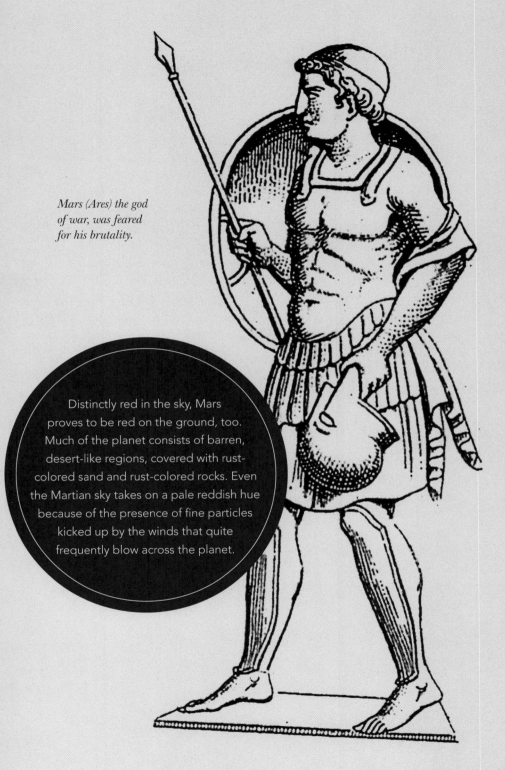

Mars (Ares) the god of war, was feared for his brutality.

Distinctly red in the sky, Mars proves to be red on the ground, too. Much of the planet consists of barren, desert-like regions, covered with rust-colored sand and rust-colored rocks. Even the Martian sky takes on a pale reddish hue because of the presence of fine particles kicked up by the winds that quite frequently blow across the planet.

Jupiter

When you see a planet shining brilliant white in the dead of night, not in the twilight of dawn or dusk, then it is bound to be Jupiter. It is visible for much of the year, shining consistently bright. The only planet to rival it in brilliance in the darkness of night is Mars. But Mars can easily be recognized by its distinctive reddish hue.

Jupiter takes about twelve years to circle around the Sun, and like all the other planets, it travels against a background of the twelve constellations of the zodiac. This means that it passes through one of the constellations every year.

By Jove

Jupiter was the king among the Roman gods. He was also called Jovis, from which our exclamation "By Jove!" comes. He was the embodiment of the Greek god Zeus and like him the god of the sky and the elements, often pictured hurling thunderbolts. The mythology surrounding the top god in the whole pantheon is quite extensive. Zeus was born the son of Cronus and Rhea, who spirited him away after birth to prevent him from being swallowed by his father (see page 117).

As a man, he overthrew Cronus and divided up the world with his brothers Poseidon and Hades. To Poseidon he gave control of the sea, and to Hades the Underworld. He kept the sky and the heavens for himself, and dwelt on Mount Olympus. Zeus was the ruler of the gods and of mortals. He was omnipotent—he could see everything and he knew everything.

The love life of Zeus was prodigious. First he married Metis, the goddess of wisdom. But it was prophesied that his first child would be wiser than its father, so he swallowed his wife and her unborn child. But before long he developed an intolerable headache, that Hephaestus cured by splitting his skull.

Out of the gaping wound leapt his offspring, not as a babe but fully grown, clad in armor and brandishing a javelin. She was Athena, the goddess of wisdom and war.

His other wives included Themis and, most notably, Hera. When married to Hera, he carried on a profusion of illicit affairs both with goddesses and mortal women, often transforming himself into other people or animals to pursue his seductions.

Jupiter is a gas giant. It has a deep atmosphere containing mainly hydrogen and helium. At the bottom of the atmosphere, the pressure is so high that it turns the hydrogen gas into liquid, creating a deep liquid hydrogen ocean. In turn, at the bottom of this ocean, the now fantastic pressure converts the hydrogen into a kind of liquid metal, rather like mercury.

Zeus (Jupiter) presided over the other gods on Mount Olympus.

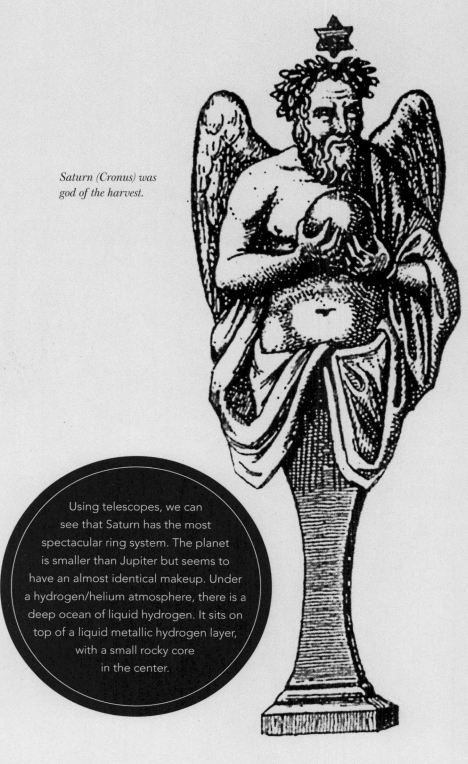

Saturn (Cronus) was god of the harvest.

Using telescopes, we can see that Saturn has the most spectacular ring system. The planet is smaller than Jupiter but seems to have an almost identical makeup. Under a hydrogen/helium atmosphere, there is a deep ocean of liquid hydrogen. It sits on top of a liquid metallic hydrogen layer, with a small rocky core in the center.

Saturn

Saturn is the most distant planet that we can easily see with the naked eye. It is nearly twice as far away as Jupiter and naturally never becomes as bright. But at its most brilliant, it still outshines all the stars in the sky except Sirius and Canopus. Unlike brilliant white Jupiter, it shines with a distinctly yellow light.

Saturn is much slower moving than Jupiter, taking nearly thirty years to circle around the Sun. It thus spends on average two years or so traveling through each of the twelve constellations.

God of the harvest

The Romans worshiped Saturn as god of the harvest, and as to be expected, he was one of the most important of the gods. One of the most riotous festivals in Roman times was the Saturnalia, which honored the god and celebrated the harvest. For the week of December 17, all work and business ceased, slaves were given temporary freedom, and presents were exchanged. We see this festival paralleled today in Christmas and New Year celebrations.

The Romans called Saturn Cronus. He was the youngest son of Uranus, god of heaven, and Gaea, goddess of the Earth. He attacked Uranus to set free his siblings, who were imprisoned (see page 118), and married his sister, Rhea. She gave birth to six children, but Cronus ate the first five because an oracle had foretold that one of them would overthrow him. When Rhea gave birth to her sixth child, Zeus, she hid him away but wrapped his clothes around a stone and presented it to Cronus, who swallowed it immediately.

Zeus was raised by the nymphs Adrasteia and Ida, daughters of the king of Crete (see page 94). When Zeus reached manhood, he vowed vengeance on his father by getting him to swallow a potion that made him vomit up the stone, and the five children that he had swallowed—three girls (Hestia, Hera, and Demeter) and two boys (Hades and Poseidon), who all became gods and goddesses. As the oracle had predicted, Zeus took over Cronus's throne and banished him to the ends of the Earth.

Uranus

Saturn was the most distant planet known to ancient astronomers. And no one suspected that there might be other planets—until 1781. In March of that year in England, a German-born musician-turned-astronomer named William Herschel spotted an object in the constellation Gemini that he first thought was a comet. But it wasn't— it was a new planet, which came to be called Uranus.

Uranus can, just, be seen with the naked eye, if you know exactly where to look. It is very slow moving among the constellations for it takes eighty-four years to circle the Sun.

The Georgian planet

Since the ancients did not know about Uranus, the planet has no mythology as such. Herschel called it the Georgian planet, in honor of his patron, King George III. The German astronomer Johann Bode suggested the name Uranus, which fitted in well with the other planets' names. In Greek mythology, Uranus was god of the heavens and one of the oldest gods, whom the Earth goddess Gaea bore and then married. They fathered the Titans, the Cyclopes, and the Hecatoncheires, monsters with a hundred hands and fifty heads.

Uranus was so ashamed of his offspring that he locked them away in the depths of the Earth. But Gaea conspired with her last-born son, Cronus, to attack Uranus. He castrated his father with a sickle and threw his genitals into the foaming sea below. From the drops of blood that spattered Gaea, the Giants and the Furies were born, and from the foam on the sea up rose Aphrodite.

Uranus is a gas giant, with a very deep atmosphere, an even deeper ocean containing water, ammonia, and methane, and a central core of rock. Its face is almost featureless. While all the other planets spin around their axis in a more or less upright position, Uranus spins on its side.

A vintage engraved illustration showing Uranus (center), with its moons Ariel, Umbriel, Titania, and Oberon.

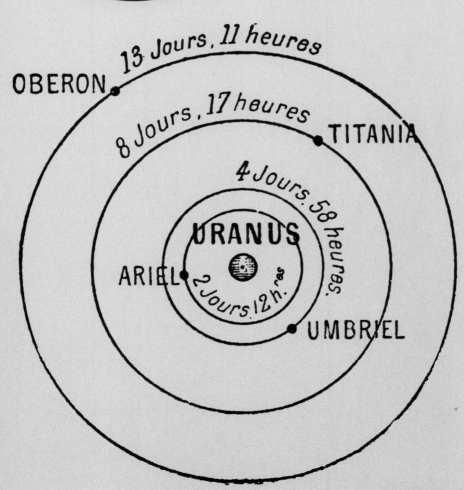

OBERON
13 Jours, 11 heures
8 Jours, 17 heures
TITANIA
4 Jours, 58 heures.
URANUS
ARIEL
2 Jours, 12 h.
UMBRIEL

Neptune

After Herschel's discovery of Uranus, astronomers found that it deviated somewhat from its predicted orbit around the Sun. So they suspected that another unknown planet must be affecting it gravitationally. Two mathematicians—John Adams in England and Urbain Leverrier in France—worked out where this planet should be. And in September 1846 a German astronomer, Johann Galle found it.

Neptune is much farther away than Uranus and is entirely beyond the reach of the naked eye. But it can be spotted through binoculars if you know where to look. Since it takes about 165 years to orbit the Sun, it hardly seems to move in the sky at all as the years go by.

The sea god

Galle proposed the name Janus for the eighth planet. Leverrier first suggested Neptune, then changed his mind and proposed to name it after himself! But Neptune soon came to be accepted. Again, strictly speaking, Neptune has no mythology because it was unknown to the ancients.

In Roman mythology, Neptune was the god of the sea. He was borrowed from Greek mythology, in which the sea god was called Poseidon. Poseidon was one of the offspring of the Titan Cronus and his sister-wife Rhea and was the brother of Zeus and Hades. One tale about Poseidon concerns a contest with the goddess Athena. The Greeks of Attica wanted to name their chief city after whoever gave mankind the most useful object. Poseidon created the horse for them, but they favored Athena's gift—the olive tree. The story explains how the city of Athens got its name.

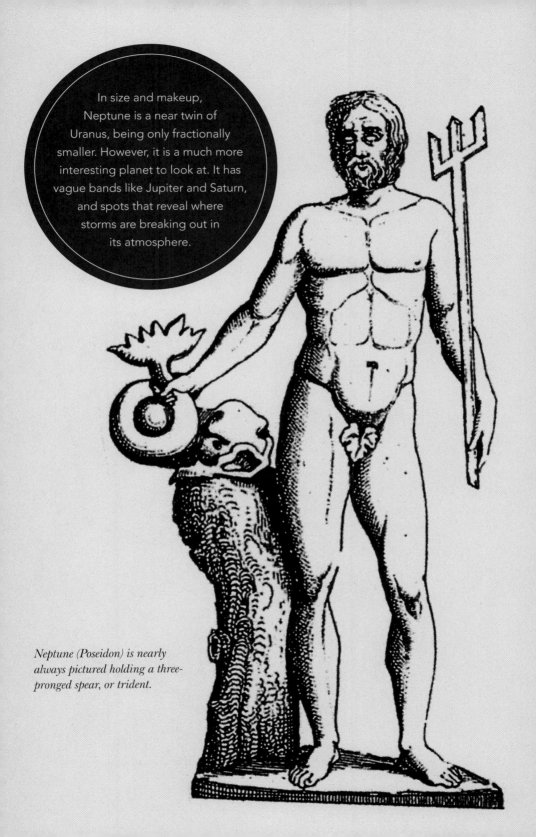

In size and makeup, Neptune is a near twin of Uranus, being only fractionally smaller. However, it is a much more interesting planet to look at. It has vague bands like Jupiter and Saturn, and spots that reveal where storms are breaking out in its atmosphere.

Neptune (Poseidon) is nearly always pictured holding a three-pronged spear, or trident.

We know less about Pluto than about any other planet because it has not yet been visited by space probes. We do know it is a deep-frozen world, made up of rock and water ice, with frozen nitrogen and other gases covering its surface. It also has a moon circling around it, called Charon.

Pluto (Hades), pictured with Cerberus, the multiheaded watchdog of the underworld.

Pluto

The discovery of Neptune in 1846 did not halt the search by astronomers for new worlds. One of the keenest was the ardent Martian enthusiast, Percival Lowell. He worked out where he thought a new ninth planet would be and set to work to find it, using his observatory at Flagstaff, Arizona. But he had no luck before he died in 1916.

Interest at the observatory in a possible new planet waned until 1929, when a young astronomer named Clyde Tombaugh joined the staff and began the hunt again. By February 1930, he had found what he was looking for, a new planet that came to be called Pluto. Though originally considered one of the solar systems "nine" planets, Pluto has been classed as a dwarf planet since 2006.

Pluto is so tiny and so remote that it shows up only as a star-like point in the most powerful telescopes. It is only about two-thirds the size of our Moon. For most of the time it is the farthest planet. But for twenty years of the 248 years it spends circling the Sun, it slips inside Neptune's orbit, and that body becomes the farthest planet. The reason for this is because Pluto's orbit is highly elliptical, or oval, which makes its distance from the Sun vary.

God of the Underworld

The Romans worshiped Pluto as god of the Underworld, borrowing him from Greek mythology, in which he was known as Hades. Hades was one of the unfortunate children of Cronus and Rhea, whom his father swallowed. But they were eventually liberated by his brother, Zeus. A classic story tells of the love of Hades for Persephone, the beautiful daughter of Demeter, goddess of agriculture. He carried her off to the Underworld (see page 96).

Glossary

ASTEROIDS Lumps of rock and metal that circle the Sun, mainly in a broad belt (band) between the orbits of Mars and Jupiter.

ASTROLOGY A belief that the heavenly bodies somehow affect human lives.

ASTRONOMY The scientific study of the heavens and the heavenly bodies.

ATMOSPHERE The layer of gases that surround a heavenly body, in particular Earth's atmosphere.

BIG BANG A mighty explosion that scientists think created the universe some 15 billion years ago.

BINARY A two-star system in which the component stars are bound to each other by gravity and rotate around each other.

CELESTIAL SPHERE An imaginary dark sphere that early astronomers thought surrounded Earth.

CIRCUMPOLAR STARS Stars close to the celestial poles, which remain visible every night.

CLUSTER A group—of stars or galaxies.

COMET A lump of icy matter that shines when it nears the Sun.

CONSTELLATION A group of bright stars that form a pattern in the sky.

COSMOS Another term for universe.

CRATER A pit in the surface of a planet, moon, or other solid heavenly body, made by a falling meteorite.

DOUBLE STAR A star that is actually two stars that are close together or appear close together in the sky. (See binary).

ECLIPSE What happens when one heavenly body moves in front of another, covering it up.

ECLIPTIC The apparent path of the Sun around the celestial sphere each year.

EQUINOXES Times of the year when the lengths of daytime and nighttime are equal.

EXTRATERRESTRIAL Not of Earth; alien.

FALLING STAR A common name for a meteor.

GALAXY A "star island" in space. Our galaxy is called the Milky Way.

GRAVITY The force every piece of matter has on every other piece.

HEAVENS The night sky.

INTERPLANETARY Between the planets.

INTERSTELLAR Between the stars.

LIGHT-YEAR The distance light travels in a year—about 6 million million miles (10 million million kilometers).

LUNAR Relating to the Moon; hence lunar eclipse—an eclipse of the Moon.

MAGNITUDE The brightness of a star. Apparent brightness is a star's brightness as it appears to us. Absolute magnitude is the star's true brightness.

METEOR A streak in the night sky made when a speck of rock or metal from outer space burns up in Earth's atmosphere.

METEORITE A piece of rock or metal from outer space that falls to the ground.

MOON A common name for a satellite of a planet.

NEBULA A cloud of gas and dust in space.

NOVA A faint star that brightens suddenly and looks like a new star.

ORBIT The path one body follows when it circles around another in space.

PHASES The different shapes of the Moon during the month.

PLANET One of nine bodies that circle in space around the Sun.

PRECESSION A slight wobbling of Earth on its axis, which slowly changes the seasons, the time of the equinoxes, and the positions of the celestial poles.

PROBE A spacecraft that leaves Earth to travel to other heavenly bodies.

PULSAR A rapidly rotating neutron star that gives off pulses of radiation.

QUASAR An object that looks like a star but is incredibly far away, and has the energy output of galaxies.

RADIO ASTRONOMY Astronomy that studies the radio waves the heavens give out.

REFLECTOR A telescope that uses mirrors to collect and focus starlight.

REFRACTOR A telescope that uses lenses to collect and focus starlight.

SATELLITE A small body that orbits around a larger one; a moon. The usual term for an artificial Earth satellite.

SHOOTING STAR A popular name for a meteor.

SOLAR Relating to the Sun; hence solar eclipse, an eclipse of the Sun.

SOLAR SYSTEM The family of the Sun, including the planets, their moons, asteroids, and comets.

STAR A huge globe of intensely hot gas that gives off enormous energy, particularly as light and heat.

STELLAR Relating to stars.

SUPERGIANT The largest kind of star.

SUPERNOVA The biggest explosion in the universe, when a massive star blows itself apart.

TERRESTRIAL Like, or relating to Earth.

UNIVERSE Everything that exists—space and all the matter it contains, such as galaxies, stars, planets, gas, and dust.

VARIABLE A star that varies in brightness.

ZODIAC An imaginary band in the heavens through which the Sun and the planets appear to travel.

GREEK ALPHABET

α	alpha	ι	iota	ρ	rho
β	beta	κ	kappa	σ	sigma
γ	gamma	λ	lambda	τ	tau
δ	delta	μ	mu	υ	upsilon
ε	epsilon	ν	nu	φ	phi
ζ	zeta	ξ	xi	χ	chi
η	eta	o	omicron	ψ	psi
ϑ	theta	π	pi	ω	omega

Index

Credits

Author Robin Kerrod is well known for his books on
astronomy and space. Among his bestselling works are
The Star Guide and the interactive *The Sky at Night*.
He also wrote *Get to Grips with Astronomy* and two series
for children entitled *The Solar System* and *Looking at Stars*.

Picture Credits